Living Terraces in Ethiopia

EASTERN AFRICA SERIES

Women's Land Rights & Privatization in Eastern Africa
Edited by BIRGIT ENGLERT
& ELIZABETH DALEY

War & the Politics of Identity in Ethiopia:
Making Enemies & Allies in the Horn of Africa
KJETIL TRONVOLL

Moving People in Ethiopia:
Development, Displacement & the State
Edited by ALULA PANKHURST
& FRANÇOIS PIGUET

Living Terraces in Ethiopia:
Konso Landscape, Culture & Development
ELIZABETH E. WATSON

Eritrea:
*A Dream Deferred**
GAIM KIBREAB

* forthcoming

Living Terraces in Ethiopia
Konso Landscape, Culture & Development

ELIZABETH E. WATSON
Lecturer in the Department of Geography
University of Cambridge

JAMES CURREY

James Currey
is an imprint of Boydell & Brewer Ltd
PO Box 9, Woodbridge, Suffolk IP12 3DF, UK
www.boydell.co.uk
and of Boydell & Brewer Inc.
668 Mt Hope Avenue, Rochester, NY 14620, USA
www.boydellandbrewer.com

First published 2009

1 2 3 4 5 13 12 11 10 09

British Library Cataloguing in Publication Data
Watson, Elizabeth E.
Living terraces in Ethiopia : Konso landscape, culture & development. -- (Eastern Africa series)
1. Konso (African people)--Agriculture. 2. Terracing--Ethiopia--History. 3. Cultural landscapes--Ethiopia.
I. Title II. Series
630.8'9935-dc22

ISBN 978-1-84701-005-6 (James Currey Hardcover)

Typeset in 10/11 pt M. Baskerville
by Long House Publishing Services, Cumbria, UK
Printed and bound in Great Britain by
CPI Antony Rowe, Chippenham and Eastbourne

For my parents

& in memory of Kahano Kalayta (1943–2007)

friend, guide & co-researcher

Hardly anything could be more isolated or more self-contained than the lives of these two walking here in the lonely hour before day, when grey shades, material and mental, are so very grey. And yet their lonely courses formed no detached design at all, but were part of the pattern in the great web of human doings then weaving in both hemispheres from the White Sea to Cape Horn.

Thomas Hardy, 1887, *The Woodlanders*

An eminent philosopher among my friends, who can dignify even your ugly furniture by lifting it into the serene light of science has shown me this pregnant little fact. Your pier-glass or extensive surface of polished steel made to be rubbed by a housemaid will be minutely and multitudinously scratched in all directions; but place now against it a lighted candle as a centre of illumination and lo! The scratches will seem to arrange themselves in a fine series of concentric circles around that little sun. It is demonstrable that the scratches are going everywhere impartially, and that it is only your candle which produces the flattering illusion of a concentric arrangement, its light falling with an exclusive optical selection.

George Eliot, 1874, *Middlemarch*

[Nature] is not a surface of materiality upon which human history is inscribed; rather history is the process wherein both people and their environments are continually bringing each other into being.

Tim Ingold, 2000, *Perception of the Environment*

Contents

List of Photographs

List of Maps, Tables & Figures

Maps

Tables

Figure

Acknowledgements

Many people and institutions have provided support, guidance and help during the research and preparation of this book.

The research was made possible by generous grants from the Economic and Social Research Council (UK), Peterhouse, Cambridge, the Nuffield Foundation, the Department of Geography, University of Cambridge and Newnham College, Cambridge. While in Ethiopia I was affiliated to the Institute of Ethiopian Studies and to the Departments of Geography and Anthropology, Addis Ababa University. I am particularly grateful for the advice and support of Alula Pankhurst, Yeraswork Admassie, Taddesse Berrisso, and Bahru Zewde.

In Cambridge, thanks go to Bill Adams, Tim Bayliss-Smith, Emma Mawdsley, Susan Owens, Sarah Radcliffe, Keith Richards, Bhaskar Vira and Ian Willis. Christopher Clapham provided detailed and much-valued advice and support. Thanks also to Helen Morales, Sheila Watts and Glen Rangwala; and to Ian Agnew and Philip Stickler who made the maps presented here.

In Konso, I am grateful for the cooperation I received from the Administration of Konso special *wereda*. I am also grateful for the support of the *Mekana Yesus* missionaries and church workers, especially Margret Hrobsjartsdottir, Benedikt Jarsonarson and Kusse Kushshu. In Konso and Addis Ababa, Korra Karra, Kebede Gelsima Kusse, Dinote Kussia, Mengistu Dinote and Gelebo Sengogo were all generous in their help and advice. In Karate, Alemitu Abebe, Aylito, Sanga Yavello and many others provided friendship and help. In Dokato, Ayana and the family of Mulu often provided a place to rest and warm hospitality. Out of town, the families of Bamalle and Kalla welcomed me and explained much to me about the lives and work of the *poqalla*s.

The people of Buso are owed particular thanks and acknowledgement for their role in this study, for allowing me to live with them and to make a study of their village. Most importantly, thanks to the family of Xalale

Xawde (Xalale, Kahana, Odiena, Gussi, Tabach, Koyiti, Xasso and Kutello) for giving me a home and becoming my family. Pacha Akicha, Toraito Xelle, Kaiyola Kaiya, Oda Kaiya, Kaso Kaiya, and Koyitti Kaiya, Koyitti Aija, Kuchulu, Gusse and Gomacho provided friendship and help with the research.

In Konso, several people worked with me and contributed in different and significant ways. Garro Kussiya was the most patient and knowledgeable research assistant and language tutor. Ongaye Oda contributed many thoughtful and critical insights. Both Garro and Ongaye made substantial contribution to the direction and nature of this work. Beyene Kailassie helped with the visits to other villages at the beginning of the work, and Halgeyo Jillo also provided assistance. In Buso village, Kahano Kalayta was a valued guide, friend and research partner. Kahano, a blind man with no formal education, literacy skills or English, had a commanding grasp of history, of dates, and of the details of life in the village. He fitted Iliffe's (1987) description of a blind 'remembrancer': someone who makes a living through recording and telling histories for people. Sometimes, after carrying out interviews together, we would spend a day or two discussing our findings, and tape these discussions. We were usually joined here by Kahano Kolobo, another knowledgeable elder. These sessions were like focus groups or seminars where people, many of whom had participated in the research, could reflect on the findings. They were opportunities to explore the material carefully and to analyse what others had told us. Some quotes from these discussions are included in what follows.

Throughout the research process, Wendy James has provided valuable advice and support. Thanks also to James Fairhead, Mats Widgren, Paula Meth, Linda McDowell, Anna Davies and Harriet Bulkeley, and to Taddesse Wolde Gossa and Dena Freeman who provided important fellowship during fieldwork in the Ethiopian South. Dick and Pauline Watson have provided unerring intellectual and emotional support, and this book would not have been possible without them. Thanks too to Rachel Watson for her good judgement and support, and to Mark Thompson for some very important help right at the end. Any mistakes, of course, are all mine.

crops. But the Konso case underlines the point that attention must also be paid to the social and cultural dimensions of indigenous knowledge. Indigenous knowledge, like all knowledge, is a product of the culture and society in which it is generated and performed. Knowledge is created, passed on, modified, organized and enacted in a particular social and cultural context. Technical know-how and access to technology mean nothing if the capacity to implement that knowledge, to maintain a set of practices and technologies, or to distribute the benefits, is absent. Extracting and transferring indigenous knowledge from one context to another will only be possible if the organizational capacity required to implement that knowledge successfully is present (even if the ecological conditions are similar). The point may seem obvious, but in the field of environmental management, the focus has often been on technique at the expense of the social and cultural dimensions that make that technique possible. An additional and related point is that, as knowledge is performed by people who are organized in particular patterns and through particular ways of managing access to and use of resources, the practice of indigenous knowledge is also political. Consideration of the power relations involved in the practice of indigenous knowledge is necessary if development organizations are to have any chance of achieving their aims of helping the poorest and most marginalized. The study of Konso is designed to illustrate how these social and cultural dimensions of knowledge and landscape production can be understood, and to explore their practical and political significance for environmental management and development. In the following sections of this introduction, I demonstrate that, although the importance of the social, cultural and political dimensions of environmental management has been understood before, the approaches to those dimensions have been limited. I argue for a more nuanced and ethnographic approach that centralizes culture, power and change over time in the analysis.

Institutional approaches to environmental management and development

In the language of development literature, the social and cultural dimensions of environmental management are usually referred to as institutions. Their practical significance is illustrated beyond doubt with reference to the example of irrigation. In Africa, hopes and investment have been placed in irrigation development as it has been seen as a way to transform environments and livelihoods for the better.[3] In the main, irrigation developments have underperformed seriously, however, and escalating costs, environment and technical difficulties have been only part of the explanation. According to Adams (1992), problems have often arisen because there has been inadequate consideration of the institutional dimensions of irrigation management. For an irrigation system to succeed, its 'operation and maintenance' must work efficiently: water pipes and furrows need to be kept clear; pumps need to be replaced; water distribution needs to be carried out in an acceptable and predictable manner, otherwise 'farmers respond to the unpredictability in water supply by ignoring recommended water use practices, over-watering while water was available and stealing water by breaking bunds or locked sluices, or bribing technical staff' (Adams, 1992: 172). Conflict between users and attacks on irrigation infrastructure lead inevitably to less efficient irrigation. Central to the success of irrigation is the institutional capacity to manage the irrigation system and its users and would-be beneficiaries.

The key role played by institutions is also seen in other kinds of environmental management. In the management of common-pool resources there is a burgeoning literature on the institutions for managing and controlling access to and use of these resources, following the seminal work of Ostrom (1990). For Ostrom, institutions allow 'a group of principals who are in an interdependent situation [to] organize and govern themselves to obtain continuing joint benefits when all face temptations to free-ride, shirk, or otherwise act opportunistically' (ibid.: 29). For scholars and development theorists, institutions have the capacity to facilitate the cooperation between people for their common good, and they sanction selfish behaviour. The key question then becomes how to optimize conditions in which the desirable institutions can flourish. If this can be done, the 'tragedy of the commons' (Hardin, 1968) can be avoided.

The interest in institutions has become widespread as it has been realized that they make an enormous difference to development outcomes. Economists such as North (1990) argue that the quality of institutions explains economic growth, or the lack of it. For North, institutions are distinct from organizations. Institutions are 'the rules of the game in a society', that 'structure incentives in human exchange, whether political, social or economic' (North, 1990: 3). They create opportunities, and some are better at this than others, explaining improved economic performance in one country or region. Other scholars prefer to use the term 'social capital' to refer to these institutional dimensions. This term was coined to refer to the 'features of social organisations, such as networks, norms and trust that facilitate action and co-operation for mutual benefit' (Putnam, 1993: 35). Putnam argued that the existence of social capital in networks of trust and cooperation reduced transaction costs and therefore generated economic growth. These approaches have been welcomed by powerful development actors. The World Bank and others have heralded social capital as the 'missing link' in development (Harriss, 1997). Formerly, attention in environment and development projects had been on technology or economic growth. The research and theories of institutions and social capital led to an awareness that the often less tangible dimensions of human development are essential to the generation of economic growth and to the distribution of its benefits.

The question for development practitioners has then become: how can social capital be built, or how is it possible to get the institutions 'right' for development (Cleaver, 2000: 361)? Answers to these questions have proved elusive, however; fostering trust and networks, and constructing rules which have acceptance and legitimacy, are complex and difficult tasks. One way to side-step the question, or to avoid being tasked with problematic social engineering and to find an answer of sorts, is to turn to indigenous institutions. Where there are established forms of indigenous environmental management, be it irrigation, terraced agriculture, rangeland management or forestry, the institutions and social capital are found to exist already and are a resource that can be built upon. After all, this is one of the main advantages that indigenous environmental management has over new imposed developments. In indigenous irrigation, for example, Wade's work has shown how there are institutions which mediate between co-users of a system, distribute water and resolve conflicts (Wade, 1988). When maintenance is required there are institutions for calling people to carry it out, organizing cooperation and for sanctioning against freeloaders. For Ostrom, there is much to learn from the institutions 'resembling neither state nor market [which] govern resource systems with reasonable degrees of success over long periods of time' (1990: 1).

In recent years, development organizations have attempted to harness existing indigenous institutions by integrating principles of customary rights, giving roles to indigenous authorities and other social institutions.[4] Generally, the indigenous institutions are considered by development organizations to be advantageous because they are thought to be adapted to the particular environment and social needs of the community. The sustainability of indigenous institutions is seen as evidenced by their presumed longevity; and they are generally considered as appropriate to be integrated into development projects because they 'can draw on the repertoire of relationships and responsibilities that already exist in the shared memory of local people' (Uphoff, 1996: x). Indigenous or customary authorities are assumed to represent the interests of their communities. Indigenous institutional dimensions are seen as a resource which can be harnessed to encourage the participation of people in a particular programme, or a means through which indigenous people can take control of their own development (Loomis, 2000). In past decades, development has been heavily criticized for cultural insensitivity, even described as a form of neo-colonialism.[5] Because these institutions are part and parcel of indigenous culture, by working with them, development practitioners are thought to be able to find new ways of achieving a development that builds on, and is compatible with, local culture (Radcliffe, 2006; Watson, 2006).

These approaches to development have been formulated with the desire to take seriously the social and cultural dimensions of environmental conservation and livelihood reproduction that make a difference to development outcomes. Some approaches are also sophisticated and attempt to grapple with the multifaceted and dynamic nature of cultural and social life. But there are also limits to the approaches outlined above, particularly when they are put into practice. When development practitioners go about identifying the institutional dimensions of environmental management, frequently a narrow and instrumental approach to institutions emerges. This is not surprising, as it is development practitioners' responsibility to find institutions that can be used to achieve certain ends. A degree of functionalism is introduced which, although understandable, is also, nonetheless, problematic. It tends to bring with it assumptions that the main or sole function of these institutions is to perform a certain environmental task, hence separating out different institutions for different aspects of production, such as tenure, division of labour, decision-making, and gender. In practice, these institutions, if they can be identified as distinct at all, are overlapping and/or linked. The very idea of identifying the relevant institutions to achieve particular ends turns the engagement with local social and cultural processes into a technocratic managerial task, which has the danger of ignoring the meanings and politics inherent in these institutions. Such an approach, that treats institutions as phenomena that can be codified quite straightforwardly and recorded, as if into a manual for environmental management, treats social and cultural life as it were something 'simple like cricket' (Baxter, 1978: 156). The analyses of Cleaver (2000) and Douglas (1987) have also shown how the institutional approach assumes that the institutions are orientated toward maximizing efficiency. It prioritizes an idea of economic (or ecological) rationality, which is highly reductive in its understanding of agency and motivation. It captures little of the way agriculture (or environmental management) is lived and practised on a day-to-day basis. It extracts the institutions from their social context and strips them of the meaning and value they have for the people who practise them. The meanings and values that the institutions are embued with are central to their production and reproduction.

The problems with the institutional approach are compounded when the institutions are also indigenous. The label 'indigenous' brings with it certain other assumptions: that they have existed in the same way for a long time; that they are environmentally sustainable; and that they serve the interests of everyone in the community. These assumptions persist despite the fact that academic studies of indigenous knowledge show the practices to be dynamic and changing (Richards, 1985). The existence of abandoned terraces demonstrates that indigenous practices can not be assumed to be sustainable. 'Indigenous' disguises the politics and differences that cross-cut any community group (Leach *et al.* 1997; Agarwal, 2001), as well as the wider regional, national and international political processes which influence practices, rights and social relations (Li, 2000). In short, although existing approaches to institutions for development and environmental management are valuable in many ways, they have certain drawbacks. They tend to extricate particular institutions

Introduction

Konso Landscape, Culture & Development

This book is about the construction of a landscape. The landscape in question is the intensive agricultural terraced landscape of Konso in south-west Ethiopia. The book focuses on the role of culture in the construction of the landscape, and explores the significance for development of the landscape itself, and the social and cultural institutions that construct and maintain it. Through this study of one landscape and one people it is hoped that the processes and connections between different aspects of people's lives and their environments will be better understood, contributing to understandings of landscape production in general, and generating insights that will be of relevance to initiatives concerned with environmental conservation and the tackling of poverty.

Konso lends itself to this study. It is an excellent example of an indigenous and intensive agricultural landscape that has been maintained for at least four hundred years, despite what can only have been enormous social changes. The Konso hills rise to a height of 2000m out of the dry Rift Valley plains, and the rugged hillsides are scored with dry stone-walled bench terraces constructed meticulously by hand. Each terrace is divided into square-ridged basins and covered in a riot of crops such as sorghum, maize, millet, *qat* (the narcotic), cotton, coffee, beans, and sweet potatoes. Trees are also grown on the terraces. There are other hilly areas in the region: some are terraced or have other soil and water conservation structures, but none is worked as intensively as Konso (Amborn, 1989). Moreover, the terraced landscape of Konso is indigenous and predates historical memory: it has been produced by the people themselves and has been cultivated continuously, sustaining lives, for a long time.

Terraced landscapes in Sub-Saharan Africa are more common than might be expected, given the stereotypical image of the African landscape as wide plains and scrub bush. Yet there are few detailed studies of African terraced landscapes, and those that exist are mostly of terraces of relatively recent provenance. These terraced landscapes have been produced, to greater or lesser levels of success, from colonial or postcolonial development policies (Tiffen *et al.* 1994; Yeraswork 2000; MacKenzie 1998; Carswell 2007). Indigenous terraced landscapes have been much less studied, and in some cases little is known about them. Grove and Sutton's (1989) list of examples is not exhaustive, but it includes twenty-three cases in West Africa; at least two in Sudan (Darfur and Kordofan); terracing in Konso and around Harar in Ethiopia; the Pare Hills and Engaruka in the Rift Valley; terracing in Rwanda, Burundi and the Kigezi area of Uganda; Ukara island in Lake Victoria; and Nyanga in Zimbawe. The terraces of Nyanga in Zimbabwe (Soper, 1996) and of Engaruka in Tanzania (Grove and Sutton, 1989) have been studied, but these have been abandoned and the studies are archaeological. Grove and Sutton conclude that hitherto research into indigenous terraced landscapes represents 'gross and misconceived offerings' (1989: 118), and that analysis has been 'mishandled at both practical and conceptual levels' (1989: 117). It seems that the striking aesthetic of terraced landscapes has hindered research endeavours: when rugged hillsides are transformed by contour bench terraces, it is as if a set of horizontal grid lines has been laid over the uneven land. The neatness is attractive and has often led researchers to interpret the terraced landscapes as physical manifestations of culture triumphing over nature, ordering it and pulling it into line, 'stamping the imprint of man (*sic*) on the earth' (Spencer and Hale, 1961: 1-2). The examinations of indigenous terraced landscapes have tended to display a degree of romanticism, an 'antiquarian and sentimental bent' (Grove and Sutton, 1989: 115), or have treated terraced landscapes as mere 'curiosity' (Netting, 1968: 4).

Most existing studies of terracing have focused almost exclusively on the material and climatological processes involving water, soil and climate. What is missing from these accounts is an understanding of the way forms of social and cultural organization work with the water, soil and climate, to produce the landscape and transform it (Parsons, 1987). Terrace construction requires huge amounts of labour to be mobilized for shifting soil and stone (Netting, 1968), and terraces require constant maintenance, as they are often damaged by rain (Guillet, 1987). The terraces may be accompanied by improvised irrigation canals, harvesting rainwater and channelling it to the crops, which require management and cooperation between field neighbours. They are usually cultivated permanently, and require frequent manuring to maintain the fertility of the soil. Terracing therefore demands 'constant and intricate attention' (Grove and Sutton, 1989: 114).

There is a gap, therefore, in understanding terraced landscapes in Africa, especially indigenous terraced landscapes, together with the role they play in livelihoods, and the ways in which they are produced and reproduced. Through a study of Konso, this book aims to go some way towards filling this gap. In doing so, it illuminates a form of landscape which has particular significance for development. Terraced landscapes are forms of intensive agriculture, which means that, because of the way the land is worked, they produce more than the land might do otherwise, and they conserve soil and water. By dividing a hillside into horizontal strips the area of cultivable land is not increased but the available resources may be used more carefully and sustainably, thus allowing permanent cultivation. The terraces prevent soil being swept down hillsides; the flat surfaces enable rainwater to infiltrate into the soil; the stone walls create micro-climates of shade and warmth reflected from the stone walls, preventing frosts (Donkin, 1979; Denevan, 2001). Blaikie and Brookfield (1987) use the words 'landesque capital' to describe structures such as this, as they are forms of investment in the land itself; the land becomes more valuable as its capability is improved or maintained, allowing the production of food and resisting degradation over time.

The value for development of such intensive agricultural systems and forms of landesque capital is clear. Surrounded by plains of less intensive agriculture, terraced landscapes have been viewed as 'islands of sustainability' or 'islands of hope' (Bebbington, 1997: 189, 191). Today there are serious concerns about food security and environmental degradation in Africa and globally. Land in Africa used to be considered abundant, but this is no longer the case. More and more, there are accounts of pressures on land and increasing numbers of landless people. Historically, understandings of change in people–environment relations have been dominated by ideas that have been characterized as 'Malthusian', which posit that population growth will lead inevitably to the over-exploitation of resources, environmental degradation, destitution and crisis. Over the past two decades many scholars have accepted the arguments put forward by Leach and Mearns (1996), and by others, that showed that Malthusian arguments were often applied erroneously to African environments. Frequently no environmental degradation has accompanied population growth; the Malthusian arguments persisted because they served certain powerful interests (ibid). However, in the current climate (meteorological, economic and political), very real pressures on lives and environments have become evident, and Malthusian arguments seem to be in ascendance again (Diamond, 2005). In this context, investigations of landscapes of intensive agriculture like that of Konso are of particular value because they show that the outcome predicted by Malthusian theories is not inevitable. It *is* possible for people to follow the contrary scenario set out by Boserup (1965), that increased populations can lead to land improvements, investments in environments, and increased or sustained food production. The scenarios set out by the Malthusian model suggest that carrying capacity (the number of people that can be supported by given natural resources) is fixed. The Konso case, together with other work on terracing by Tiffen *et al.* (1994) and Carswell (2007), shows that carrying capacity depends on how those resources are used and by whom. Intensification through terracing is not only possible and valuable, but in today's world it may be imperative. The questions then become, how can intensification be produced and sustained? Does it have a cost, and if so, for whom? What are its limits? These are the questions that are explored further through the analysis of Konso.

While terraced landscapes in general have significance for development, landscapes such as Konso have further value because of their indigeneity. Indigenous terraced landscapes are produced out of the application of indigenous technological or environmental knowledge.[1] Until the 1970s and 1980s, these forms of knowledge were frequently denigrated as inefficient, unscientific, chaotic, undisciplined, and, often, harmful to the environment. Development was equated with a particular brand of modernization in which 'Western' scientists and 'educated' administrators were seen as having the 'modern' knowledge, technology, forethought and sensibility to manage the environment and to arrest any advancing degradation. Such modern development projects often failed, however, and frequently resulted in environmental damage. At the same time, research showed that indigenous knowledge often provides ingenious solutions to difficult problems and is the product of generations of experimentation and experience in particular environments (Richards, 1985; Fairhead, 1992; de Boef *et al.* 1993; and Reij *et al.* 1996). The practices that result from the application of indigenous knowledge can be much more effective and efficient than previously thought; they make inventive use of available and renewable resources, making them environmentally sustainable. Soil conservation or irrigation structures, for example, made from local resources are easier to replace than those manufactured industrially. The techniques are also often risk averse: they focus on sustaining livelihoods and environments over time, instead of concentrating on producing surplus.

The 'discovery' of the value of indigenous knowledge and environmental practices led researchers and development organizations to try to learn from and to support indigenous knowledge. In an approach sometimes characterized as 'small is beautiful'[2], development practitioners and activists have stressed that sustainable solutions to world environment and development problems may be found in these practices. As a consequence, attempts are being made to transfer indigenous knowledge and techniques from one situation to another:

> It is quite apparent that indigenous innovations, which are found to be effective in one part of the globe, can be equally effective when made available to populations in similar ecological conditions in other parts of the world. The documentation of the vast amount of unrecorded, often rapidly disappearing indigenous knowledge could provide the basis for many effective development interventions, if this knowledge could be shared (Warren, Slikkerveer and Brokensha, 1995: xvii).

The present study follows this direction: much can be learnt of value for development from Konso where local people combine multiple soil and water conservation techniques to maintain soil fertility and raise food for particular purposes from their context, when that context may be an inherent and central part of the institutions themselves. Such approaches rarely take account of the power politics implicated in those institutions and institutional practices, or the meanings that the institutions have for the people who replicate them. All too often the approach crystallizes a particular moment in time, paying little heed to the history of those institutions or the ways they are changing.

The concept of institutions is useful as a shorthand for the multiple social and cultural dimensions that are vital to people-environment relations, to landscape production and to development outcomes. Policy-makers need shorthands and simplifications in order to be able to communicate ideas and generate action. But the assumptions contained within simplifications can have dangerous consequences. If adverse consequences are to be avoided, those wishing to harness the power of institutions for development must develop an awareness of varying dynamics that are contained within the simple term. Empirical studies, like that of Konso presented here, can improve understandings of the nature of indigenous institutions and their usefulness for development, but an approach that embraces them in all their complexity, politics and dynamism is required.

A broader, more encompassing approach to institutions is suggested by Leach *et al.* (1997), who dismiss the need to maintain North's distinction between institutions and organizations. Their more practice-based approach defines institutions as 'regularized patterns of behaviour that emerge from underlying structures or sets of "rules in use"', where '"rules" are constantly made and remade through people's practices' (1997: 27). Institutions include forms of organization, conventional knowledge, 'regularized practices', customary rules, rights and responsibilities.[6] They determine legitimate and illegitimate use and practice. When applied to the environment they define who can use which resources, how much they can use, and the type and manner of that use. Importantly, this conceptualization includes informal as well as formal institutions. The relations between people, development and their environment, are produced out of a matrix of these different overlapping and intersecting institutional arrangements, operating at different scales (local, regional, national, international).

Leach, Mearns and Scoones' approach derives much from structuration theory: structures constrain and enable certain forms of human action, but human agents can also transform those structures over time (Giddens, 1984). The practice-based approach to institutions also takes much from Foucauldian approaches to discourse and power as developed by theorists like Hajer (1995). Here, institutions are structures of power. Once established they may have a certain degree of staying power, but they have force 'only to the extent that they are constantly reproduced in actual practices' (Hajer, 1995: 58). In order to understand which practices are considered legitimate, and how these change over time, it is necessary to look to the discourses which frame and give meaning to those practices. The definition of discourse that is drawn upon goes beyond mere verbal expression and written documentation to practice itself: 'Discourse is here defined as a specific ensemble of ideas, concepts and conceptualizations that are produced, reproduced and transformed in a particular set of practices and through which meaning is given to social and cultural realities' (ibid.: 44).

These approaches centralize issues of power and change in the analysis. Discourse is related to a subject's position and power. The world is made up of many different discourses, contesting, competing and negotiating with each other. It is through the promotion of a different discourse and its acceptance that different social groups can promote a certain way of doing things. By constructing and identifying with a particular discourse an individual's very sense of self and position in the world is articulated in particular contexts. Discourse is therefore a practical and personal matter, and not some over-arching, independent, ephemeral phenomenon. The engagement of discourse is a lived reality, and one through which different groups, and even individuals, can play out different strategies. Those strategies are limited by the institutional context within which a person or group find her/his/themselves, but that institutional context can also be changed through her/his/their actions.

Leach, Mearns and Scoones' approach to institutions is attractive and it has been influential, but it contains within it enormous levels of complexity. Analysing the institutions that operate in any situation involves examining the ways in which multiple institutions, formal and informal, intersect, exploring their consequences for different individuals and groups, and investigating the way they change over time. Such an analysis involves exploring the ways ideas, attitudes, values, meanings and structures exist, intersect, and the ways they influence behaviour, identities, positions and wider structures. This approach is followed here, but it is also argued that carrying out such a study essentially involves studying society and culture with all its multiple dimensions, politics and change over time. If a recent definition of culture is examined, it is evident that there is a great deal of similarity between the processes that Leach, Mearns and Scoones have been describing and the structures and processes of 'culture':

> Culture refers to loosely structured clusters of practices based on values and meanings that are shared, but also contested. These practices are interrelated and can produce sometimes wide-ranging or long-term effects. Cultures are emergent and collective, but only loosely coordinated and internally fractured. They involve a multitude of human actions guided in a very broad, general sense by the (sometimes enabling, sometimes oppressive) presence of the past in the form of institutions, rules and regulations, the built environment, structured inequalities, structures of power, accepted ways of acting, interpreting, and challenging these ways of relating (Duncan and Duncan, 2004: 394).

Identifying the institutions that are involved in environmental management and central to the production of landscapes therefore involves studying a culture and its engagement with its environment. Here culture can be summarized as 'the meanings that people create, and which create people, as members of societies' (Hannerz, 1992: 3, cited in Östberg and Reij, 1998: 1350). Östberg and Reij make the point that culture is not a static, rarified and rule-bound: 'cultures are in the plural and they are changing' (ibid.). There is no short cut to such a study: using the label of 'institutions' provides a useful bridge to development concerns and literature, but it does not make the study of the complexities of culture and its engagement with environment any easier to achieve. An examination of these institutional processes involves all the challenges inherent to an ethnographic study. In order to take account of change over time, it must also be a history.

In the context of rural agriculture in the global south, studies of culture and its direct relationship to environmental management and landscape have been scarce in recent years, as researchers may have been afraid of being accused of resorting to crude environmental determinism or functionalism. There are of course much older approaches to landscape and culture from geography and anthropology. For Sauer, landscape was both 'natural' and 'cultural' in that each landscape was distinct because of the way the material reality was transformed by the culture of the people who lived there (Sauer, 1963 [1925]). Culture here was viewed holistically and there was little concern for the politics or power relations implicated within it (Mitchell, 2002). Work on terracing has not completely ignored culture either. For Netting (1968), the terraced landscape of Kofyar in Nigeria was produced out of a 'cultural ecology', in which culture is viewed as an adaptation that functions to reproduce the landscape and forms of subsistence. For Netting, it is the researcher's task to identify the aspects of culture that are 'relevant' to the production of the landscape. These aspects of culture will also endure because they manifest a level of efficiency and effectiveness with regard to the reproduction of the landscape and livelihoods. Even aspects of personality and attitudes seen to be exhibited by a particular culture may be explained in relation to ecological factors. There is a closed loop, therefore, including the material ecology and certain aspects of culture. Other aspects of culture are considered superfluous and irrelevant to the account of people–environment relations. There is no attention to the power relations entailed within culture; the cultural group is homogeneous and functioning. The aim of the present study is to retain a focus on culture and its relation to the landscape, and to understand the way culture functions environmentally, without resorting to a functionalism that ignores variation, power dynamics and inequalities, and change over time. The aim is also to study Konso culture and its relationship to landscape without losing an engagement with the issues that concern development theorists and practitioners. In the next section I outline the approach employed here.

Landscape as cultural process

The approach adopted here builds on the practice-based approach of Leach, Mearns and Scoones (1997), and views Konso landscape as part of a cultural process. The title 'cultural process' is adapted from Berry's work (1989a, 1989b, 1992), that describes access to and use of agricultural resources as a 'social process'. For Berry, productive resources 'include not only material and financial means of production – land, labour, fixed and working capital – but also social relations and knowledge which may be employed in processes of production' (1989a: 142). Her work therefore addresses the human and institutional aspects of production described above. The advantages of Berry's approach lie in the way she shows that different aspects of production are interconnected with each other *and* with social identity and group membership. Rights to land, for example, depend on social identity, status and group membership. Rights to and use of land may also depend on having access to labour to cultivate it, which also depend on social identity, status and group membership. Social identity and status determine the kinds of labour that are thought appropriate for a particular person, the kinds of labour and knowledge that are learnt, how much labour is expected of a person, and the benefits that are derived from it (Katz, 1991; Berry, 1989b). Access to resources, labour and knowledge, therefore, all depend on the structures of social identity: as well as membership of a particular ethnic group or lineage, gender, race, class, generation, or marital status may all be important. [7] The approach is holistic, in that it looks at the interconnectedness of social, practical, economic and political dimensions of production. All these are changing, but they are not necessarily part of an environmentally-adapted system.

To Berry's basic approach, this study adds an emphasis on the importance of the cultural dimensions of meaning that are attached to the processes outlined above. The interconnected access to resources, agricultural practices and social identities are all replete with meanings through which they are valued, understood, experienced, performed and accepted or rejected over time. There are many possible ways of approaching and examining these cultural dimensions. Here, the concept of discourse, when defined broadly as outlined above, is seen as invaluable because it provides a way to understand how meaning is created and its significance. The concept draws attention to the consequences for different groups and individuals of different meanings, ideas, narratives and signifying practices. It explores the power politics inherent in any arrangement and set of practices. It also provides a way to explore change and continuity, over time, as well as (in certain conceptions of discourse) to explore the relationship of a particular individual to the wider discursive structures and the opportunities they afford and/or prevent.

society. Some followed the 'modern' set of beliefs and practices and managed their land and labour according to *Derg* policy specifications; others continued to respect their *poqallas* (if secretly) and continued to manage land and labour according to the customary practices. The two discourses existed in parallel. What is striking is that although Konso experienced dramatic changes in this modern period, the orthodox discourse (and the practices of land and labour management that it encompassed) was not defeated in the discursive struggle that ensued. Konso customary practices have demonstrated an enormous level of resilience and autonomy in the face of changes brought by the state and globalized religion.

The study of the two discourses in Konso, the orthodoxy and the heterodoxy, shows that Konso environment and landscape has been, and continues to be, produced out of complex processes that are social, political, cultural and spiritual, as well as biological, technical, economic and climatological. Despite radical changes in religious ideas and ritual practices, the ritual and spiritual dimensions of the more practical activities and forms of organization remain central. It would be possible to conclude that the sacred domain is central to the mundane, if this did not imply that the two existed in discrete spheres. The more ordinary practical activities are shown here to be in some ways themselves sacred. The Konso landscapes of terraces are also therefore landscapes of faith, albeit, over time, of different faiths.

In Chapter Seven changes that have happened in Konso in the post-1991 period are explored. The post-1991 EPRDF government promoted a programme of ethnic decentralization and championed the rights of minorities to practise their own culture. This context has given new impetus to the orthodoxy, and the customary beliefs and patterns of land and labour management have experienced, to some degree, a resurgence. This 'retraditionalization', to use Chabal and Daloz's (1999) term, has had limitations, however, largely owing to the politics implicit to the customary practices. Exploring the politicization of the *poqallas* and other aspects of the customary provides insights for development organizations that might wish to form partnerships with indigenous institutions. While the processes described as the heterodox discourse have been discussed as a 'modernization' of Konso agriculture, the processes described in Chapter Seven are described as a 'postmodernization'. The label is useful for exploring the way the period is one characterized by uncertainty and by multiple identities. Conceptual categories that were formerly opposed are now being brought together and there are alliances between previously opposed groups as well as new points of conflict. In these new developments, religious ideas and discourses are only one factor; the new political situation and new economic opportunities represent novel challenges.

In the Conclusion, we return to the landscape more explicitly, to examine the extent to which the turbulent cultural history that has been described is evident in changes in the form of the landscape. Some further conclusions are also made about the extent to which culture, in the form of these indigenous institutions, can be harnessed by organizations for environment and development purposes.

In summary, this book explores the way institutions are central to the production of landscapes and to the practice of indigenous knowledge. Institutions are structures of power, and they are implicated in social, political, cultural, spiritual and economic life. They change over time as a consequence of endogenous and exogenous forces of change. Approaches to institutions have attempted to grapple with this complexity. The list of types of institutions as 'regularized practices' and 'sets of rules of use' gives the impression that it would be quite easy to identify and work with the institutions underlying livelihoods and landscape production. In practice, understanding the overlapping and multi-dimensional cultural structures of power that are in flux is much more complicated. History, social anthropology and geography have for decades struggled with attempts to understand and describe these aspects of life. Development practitioners and theorists wishing to understand the institutional dimensions of environmental management can employ the same ideas and approaches to good affect. Here, readings of Foucault's concept of discourse and Bourdieu's orthodoxy and heterodoxy, provide a means to engage with institutions as structures of power, while retaining aspects of meaning and experience and local colour. They also provide a way to capture the nature of the institutions and the way they change over time, as institutions, like the cultures which they make up 'do not', as Clifford wrote, 'hold still for their portraits' (1986: 10).

The aim is to contribute some insights that are of relevance to development theory and practice, and to those activities that aim to improve the lot of people who are receiving a disproportionately low share of the world's wealth, and who suffer routinely from problems such as food shortages, disease or poor medical care. The study demonstrates that indigenous agricultural systems such as that of the Konso are valuable, but that it is difficult to abstract and employ instrumentally those techniques or the institutions in which they are embedded for development ends. But this book is not an extensive study of development projects themselves, although one or two projects are discussed towards the end of it. It is more concerned to understand the processes that are taking place on the ground that have been in existence much longer than any development initiative, and are likely to continue (albeit changed and changing) after such development initiatives have been completed. If development initiatives are to be successful they must have good understandings of the contexts in which they work, and studies that look in depth at these contexts can critique some of the assumptions that development projects employ in their work and practices. One of the main motives for writing this book is to provide a meaningful study of the Konso people and their lives, and to explore the way in which they have changed over time. It is hoped that, as well as grounding some of the debates about indigenous agriculture and institutions in real situations, the book might be of interest and use to the Konso people, who may debate and contest the account presented here. Although this book looks in depth at one place and one people, it is also hoped that the approach and the insights will have relevance for other peoples, processes and places.

Provenance of this research

Before continuing to describe Konso landscape, culture and history some background to the study and information on the research methodology is useful. I keep this section brief, while at the same time aiming to elucidate the experiences and processes which often remain undisclosed, but which can also have a strong influence on the research process.

The research emerged from twin interests in indigenous agricultural knowledge and social and cultural change in Africa. While, on the one hand, I was fascinated by the indigenous irrigation and soil and water conservation systems and the way they supported livelihoods, I was equally fascinated by the way those who used the indigenous systems, especially the young people, were often much more interested in the new, in technologies, cultural ideas and practices that came from 'outside'. These two interests came together in the research project exploring the role of culture and cultural practices in indigenous environmental management. More specifically the research centred on the question of 'if the practice of indigenous agricultural knowledge is so embedded in social and cultural institutions, what happens to those practices in times of extreme social and cultural change?' My experiences up to then had led me to believe that social and cultural change was the norm rather than the exception for Africa, even in supposedly remote rural Africa. I also felt that researchers still had much to do to meet the challenge of coming to grips with the dynamics of societal and cultural change. Since Hannerz's call for the 'creolised' cultures of Africa to be given more attention (Hannerz, 1987), there had been more work of this kind, but there was still much to be done in terms of studying contemporary African modernities. There is also little work that examined questions related to contemporary culture and change, which also explored the practical developmental questions related to everyday livelihoods. Studying change, like studying culture, is not easy, however. I should confess to hoping, rather naively, that by studying cultural change and landscape together, the cultural changes might be traced in the landscape, almost as if they had been mapped onto it. As this book shows, it is not always that simple.

Konso was selected for the study as it is the northern-most case of a set of intensive and specialised indigenous agricultural systems, each of which has a long history, that were written about in an edition of the journal *Azania* in 1989. Very little was known about the agriculture of Konso at

Photo 0.1 *Konso landscape in the dry season (looking from Dokato towards Buso and Gamole)*

the time, however, even though it was the only case of its kind in Ethiopia. In addition, when this research was begun in the mid-1990s, Ethiopia had just been through a change of government and there was a favourable research climate.

Fieldwork started in 1995, when I spent 21 months in Konso as part of my graduate studies. Much of the detailed quantitative material on land and labour management presented here is taken from a survey that was carried out in 1996. The detail contained in this survey has proved invaluable for demonstrating that the *poqalla*s, the lineage leaders and 'priests', play a major role in the production of the landscape. It also shows that their role is not just some theoretical cultural ideal discussed in interviews, but was present in people's everyday practices. During the first period of fieldwork, I also spent time learning the language, *Konsiña*, or *Afa Xonso* as it is known locally (see Watson, 2004 for an account of this process), and carrying out participant observation. I interviewed key informants, collected oral histories, attended rituals, followed events, and generally attempted to gain as full a picture as possible of Konso culture, its history and the processes through which land and labour are managed and produce the landscape. The first lengthy period of fieldwork was followed up by shorter periods in 2002, 2005 and 2008. As well as extending and deepening the understanding of Konso culture and landscape, the later periods of fieldwork explored in more depth the impact of the ethnic federalization process, the engagement of Konso

Photo 0.2 *Inside Buso village*

people with the post-1991 Ethiopian state, and the processes through which *poqallas* have reasserted their power and position.

Much of the research presented here centres on the village of Buso, some three miles from the administrative centre of Konso special *wereda* (see Chapter Two). Early in my time working in Konso, Buso village took me in and the family of Xalale Xawde in Buso gave me a home. Buso was one of the main villages where C.R. Hallpike had carried out his anthropological fieldwork nearly thirty years before. I had rather hoped to stay in another village, but early in my time in Konso I had been from village to village talking to Peasant Association Committee members[10] and to elders, trying to find a place to live and research. All seemed dubious about the idea: most said that they had nothing against the idea in theory, but that they couldn't see how they could find any space for me.[11] Then Xalale Xawde, a man whom I'd met in another context, came to see me and said he'd heard I was looking for somewhere to live and that I could stay with him and his family. Thus, in some ways, Buso selected me rather than the other way round. As one of my main interests was in change over time, working in a place where there was some historical record seemed to be potentially an advantage. Hallpike's 1972 ethnography has been drawn on in this research. He also studied the connections between the moral social world and the natural, although he was less interested in the nature and form of that natural world itself. The present focus on the social institutions that play a role in the production and the reproduction of the

Photo 0.3 *People drinking* chagga *in our home*

landscape provides a more practical and grounded perspective on Konso lives and experiences than the work of Hallpike, and gives a different perspective on some of the institutions that he studied. In addition, he viewed the moral values of the social world as informing the 'rules' of the 'game' of life, which implies a degree of strategizing on the part of actors, and a rigidity to the cultural forms – at least for a time – which is avoided here.

Living in Konso

In Buso, I lived with Xalale, his wife Kahana, and five children ranging at the time from the age of about sixteen to two. The compound was on two levels: the houses were on the upper level (*oyta*), and the granary with grinding stones beneath; the stall where our cow, Kurati, was kept was also on the lower level (the *arxatta*). There were two houses; Xalale, Kahana and the children slept in one, and I, the other. Staying with the family was invaluable for allowing the participant observation and language learning. I spent the days working with the family in the fields, mostly weeding, collecting fodder for the cow, occasionally making *chagga* (the local grain beer) with Kahana. Sometimes I stayed in the compound shelling beans that the elder daughter would bring back from the fields for supper, or attempting to spin cotton. I did not find these activities or living in the village easy. The work could be difficult, I had very little privacy, and children especially took great amusement in coming to watch me. The sanitation in the village was poor, and I tried to partake in the main adult pastime of both men and women: drinking *chagga* (beer) early in the morning, for lunch and in the evening. The Konso consider *chagga* to be 'their number one food'[12] and otherwise they usually cook just once a day, in the evening. A woman would often brew *chagga* once a week and sell it to other people in the neighbourhood, who would come and sit in her compound. Thus, despite my reservations about such frequent and regular alcohol consumption, drinking *chagga* with either Kahana or Xalale introduced me to many people and to other homesteads.

Becoming part of this family gave me a Konso identity: people extended the family's name and clan affiliation and membership of the neighbourhood to me. I became thought of as the daughter of Xawde family, woman of Sawdata clan, of Kolalta neighbourhood in Buso village. This had repercussions throughout my work, as it enabled me to fit into networks of kin and clan affiliations. I could respond adequately to questions on identity that were asked after greetings, and which previously had just emphasized my 'outsider-ness'. Now many people referred to me as their clan-sister (*inenda*), or as their daughter-in-law (*ashumta*) if someone of their family had married a member of Sawdata clan. All this was carried out partly tongue-in-cheek, but it served to place me in the Konso social landscape. Even doing fieldwork now people being interviewed often ask if I am still making *chagga*, or carrying manure to the fields (the answer is invariably no) and it is clear that the time I spent living and working with the family had much greater significance for the research than just being an important process of learning by doing.

Throughout my fieldwork, I also kept a room where I could rest and work in the administrative centre, on the compound of the Protestant Church or in one of the small hotels. The missionaries and staff of the church provided generous support in the early period of this work and asked for little in return. When I stayed in the mission, I was part of a different community. My neighbours were other young single women who were staying there while they studied in secondary school or worked in the health centre, mission or government offices. Their friendship provided personal support and guidance, and also provided insights into the experiences and perspectives of Protestant Christians. I was afraid, however, that living on the church compound would bias whom I talked to and the perceptions people had of me: hence the move and why the identification with Buso village and Xalale's family was so valuable.

As well as working in Buso, research has also been carried out further afield and is drawn on throughout this book. I interviewed regional *poqalla*s who generally lived outside villages throughout Konso. I visited other villages and made comparisons with Buso. I also followed events and rituals in the wider Konso area. All these research activities have helped to put the Buso case study in its wider context.

The Konso people I have worked with during this research have always shown me an enormous level of warmth and encouragement. They accepted me and my work at face value. They participated in the research openly and with good humour, weighed up questions carefully and thoughtfully, and shared their histories, experiences and ideas. I hope they feel that their practices and experiences are represented fairly here, and that the analysis adds something useful to understandings of Konso lives and landscapes, contemporary and historical.

Notes

1 Sillitoe *et al.* (2002, 9) review the different approaches to these kinds of knowledge, also known as local knowledge, rural people's knowledge, insider knowledge, indigenous technical knowledge, traditional ecological knowledge, ethno-science, people's science and folk knowledge.
2 After Schumacher, 1973.
3 Irrigation development is given high priority in Ethiopia: in 2002 the Ethiopian Prime Minister, Meles Zenawi, singled out the lack of investment in the development of irrigation in the country as one of the reasons for the recurrence of food shortages, and he identified a need to develop irrigation in the future in order to meet food needs.
4 For examples see Watson (2003), Black and Watson (2006), Watson (2006).
5 This critique is associated with the post-development and anti-development schools of thought (see Nederveen Pierterse 2001 for review).
6 See Watson (2004) and Watson *et al.* (1999) for review.
7 Even the supposedly most fixed social identities change over time. For Berry, social identity is both 'ascribed and achieved' (1989b: 42). Membership of different groups can be achieved (even across ethnic boundaries; Spear and Waller, 1993) by participation in the appropriate rituals, or by the acceptance of the authority of the dominant group of landowners (Schoffeleers, 1979). Individuals can also 'advance their status (or even alter their identity) by personal achievement and the adroit use of wealth' (Berry, 1989b: 42).
8 Some of these authors do not use the term discourse in their work, but instead focus on the production of meaning through different kinds of practices (see for review, Moore, 1999). I believe that this work is not incompatible with the broad concept of discourse, outlined above. I use the term discourse explicitly in order to make a connection between the different kinds of discourses and particularly with the literature on discourses and narratives and the way in which they shape the production of landscapes.
9 The term 'orthodoxy' or 'orthodox' should not be confused with the Ethiopian Orthodox Church.
10 The Peasant Association Committee is the lowest level of state administration. Almost every village has its own elected group of people who make up this committee.
11 There are many reasons why people may have been wary of having me live in their village. They probably thought it odd that a woman on her own would want to move into a village. Space in the villages is also tight.
12 This was Ongaye Oda's phrase. The alcohol content of *chagga* varies enormously.

One

Konso Intensive Indigenous Agriculture

The Konso people live on and around a small range of mountains, some 600 km south of Addis Ababa, in the Rift Valley of Ethiopia (see Map 1.1). These mountains rise to a height of 2,500 m, and from above 700 m they are scored with neat, dry stone terraces (Amborn, 1989). To the outsider, the area is distinctive in many ways: the terraced landscape marks it out as different from the hot lowlands that surround the mountains, and from the other cultivated mountains in the area. In the area marked on the map, they also speak their own language, *Afa Xonso*, an Eastern Cushitic language. It has some links to *Oromiffa* spoken by the Boran and the Guji to the east, and other links to the Dirashe to the north-west, but the languages are not mutually comprehensible. The Konso are also recognizable by their style of dress: women often wear thick home-spun cotton skirts which are gathered and folded into two layers; the men sometimes wear home-spun and woven cotton shorts, often striped and brightly coloured.

Konso villages are nucleated, walled and extremely dense and compact. The density of the settlements and the way in which they are criss-crossed by many narrow and labyrinthine paths led Hallpike (1972) and others, Sutton (1989), to describe these settlements as towns. Though in many ways this word is appropriate, it is less confusing to describe them as villages, distinguishing them from the more urban recent settlements of, for example, Fasha and Karate, both of which have grown up around market places. Karate is the main town, and the location of government offices. The towns provide other services such as schools, churches and health centres. They are also home to many Konso (though mainly those who have formal employment) who have moved there to live, together with traders, school teachers and others who work in the town. The population of Karate town swells substantially during the day as people come to visit the health centre, school, court, church and mission, or market, and the atmosphere is quite different from that in the villages.

The villages are walled, and many of them are situated on easily

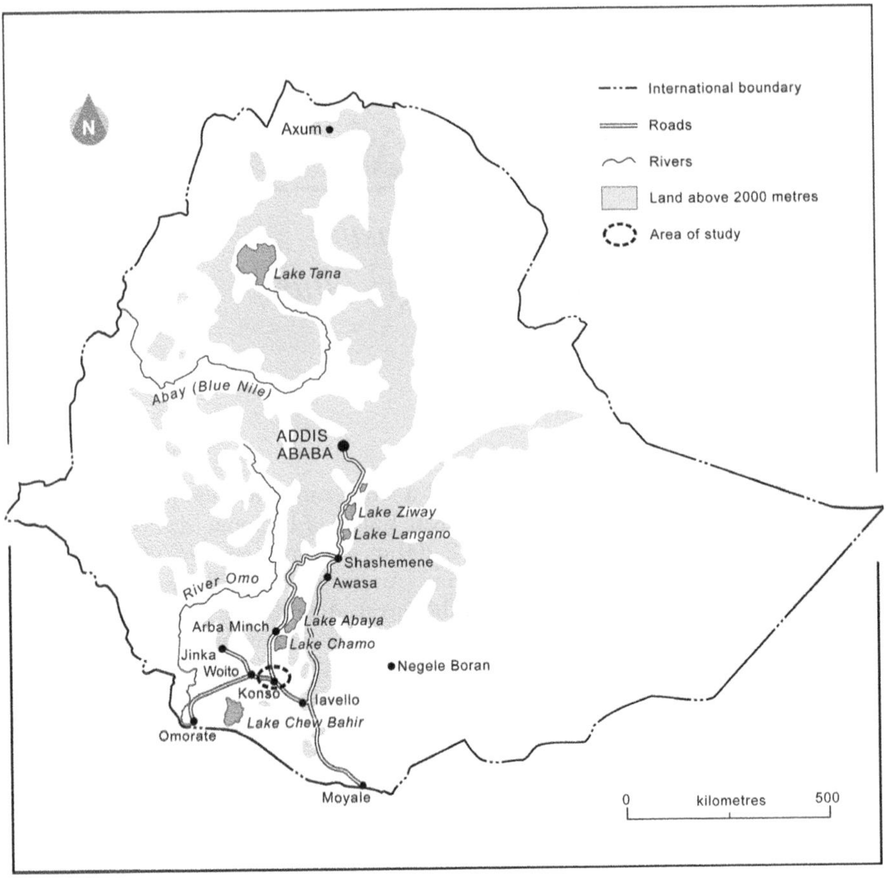

Map 1.1 *The location of Konso in Ethiopia*

Map 1.2 *Konso and approximate locations of neighbouring groups*

Photo 1.1 *Labyrinthine pathways in Buso village*

defensible sites such as the top of hills or hillside spurs. This is said to be a safeguard against aggression from other Konso villages, and also from attacks from other neighbouring ethnic groups, for example the Boran, Guji and the Tsemai, with whom relations have sometimes been hostile (Hallpike, 1972; see Map 1.2). According to the Konso, the walls were also built to keep out wild animals such as lions and hyenas, which used to be common and are said to have terrorized people and their animals. In Kulme and Turo areas (see Map 1.3), the villages are not walled and the homesteads are dispersed.

The landscape of Konso is thus of hillsides scored with terraces, with the occasional nucleated settlement. The topography and locations of the villages can be seen in Map 1.3.

Early perspectives on Konso

It is difficult to know the age of agriculture in Konso. The work of Amborn (1988; 1989) and Hallpike (1972) suggest from oral histories that it is at least four hundred years old. Ehret (2002: 134, 219) uses linguistic evidence to suggest that intensive agriculture in the region could date back as far as two millennia. As the fields are constantly worked and re-worked it is difficult to obtain clarification on age from archaeological analysis of the fields; the earliest known written accounts of Konso date back only about one hundred years.

Photo 1.4 *Terraced hillside during the dry season. Square basins, drainage ditches and* oypata *trees are visible.*

Photo 1.5 *Large working party with women and children carrying soil and men building walls*

historical data which may inform our understanding of the area. The surrounding valleys are no longer replete with game, and the tsetse fly no longer devastates passing domestic animals, suggesting a progressive reduction in scrub. This suggests that the mountainous area provided some refuge from attacks of wild animals and insect-borne diseases in the past, although malaria is still a serious problem for humans in the lowlands today. It is easy to allow a fear of sounding 'sentimental' to reduce an account of the area to only a list of techniques and practices, and to silence any sense of appreciation of the effort that goes into producing this living landscape. Terraced landscapes do not appear overnight, and need to be constantly maintained.

Integrated, specialized, intensive agriculture

Sutton describes Konso as a 'finely balanced and integrated system of specialised techniques' (Sutton, 1989: 106). These techniques represent a transformation of the environment in order that the people are not entirely at the 'mercy' of the rainfall and soil fertility. They build soil and water conservation structures which prevent soil erosion, improve the quality of the soil, collect and redirect rain water onto the fields, and maximize the use of rain water.

The rain falls in two seasons: the largest rains, known as *sorora*, fall between March and May, the main cultivation season. Smaller rains, known as *hagaya*, fall between September and November. The period of December and January is very dry, sometimes with no rain at all, and is known as *pona*. In this time the fields are prepared so that they are ready for sowing at the first sign of rain in February–March.

The mean annual rainfall is approximately 520 mm, calculated from the years when consistent rainfall data were available. Of the consistent data available, the minimum yearly rainfall was 278 mm and the maximum was 870 mm, demonstrating the unreliability of rain. The rainfall for each month shows a great deal of variation (with standard deviation from the mean of each month averaging 76 per cent over 12 months). The rain is often too little, and frequently falls at the 'wrong time'.

In Konso the terraces are dry-stone walled bench terraces. The walls are constructed by carefully fitting together large and small stones depending on their availability. The stones are collected in the field that is being terraced itself, and, in many areas of Konso, they are usually abundant. No mortar is used. Amborn (1989: 74–5) describes nine types of terrace walls that vary in sophistication: some are merely a pile of small stones; others are two rows of larger stones side-by-side (sometimes with small stones in between); other structures have outlet channels built into them, to allow excess water to drain, thus preventing flood damage. The height of the terraces varies, depending on the terrain; heights of 0.5 to 1.5 m are preferred. However, much higher walls occur, and those defensive walls around towns, which are built by the same techniques, can reach up to 10m (Amborn, 1989). For construction, the soil is cut away to make a perpendicular face, and then a wall of stones is built against it. The foundations are usually shallow, around 1025 cm (Amborn, ibid.). The earth is then moved from above the terrace and used to level the lower area. While building the walls by placing stones carefully is exclusively the work of men, men and women sometimes engage on terrace building together, and then women move the soil by collecting it up in long wooden bowls, carved from trees, called *toma* (see Photos 1.2–1.5).

The Konso call the terrace walls *kawwata,* but the name of the terrace itself is *xeeranta.* The terraces prevent soil erosion and maximize water retention and hence infiltration. Square basins (*kolba*) are also constructed on the terraces by making earthen ridges (or 'risers') of 10–20 cm high (*mona* or *korayita*). These further retain water and improve infiltration (Photo 1.6). The ridges are covered by sorghum stalks and roots, which are dug up after harvest and placed there, with clods of earth clinging to the roots. These help to form the sides of the ridges.

The terraces are combined with other water harvesting or water control structures that are built from stone and other locally available materials. Small ridges and terrace walls are made to redirect rainfall from roads, pathways and occasional streams onto the fields (Photo 1.7). The water produced by even the smallest spring is captured and redirected. Small canals are built from stone and mud banks to take water from flooding rivers (Photo 1.8). Stone dams are occasionally built across small streams to raise the level and enable water to be carried to nearby fields (Photo 1.9). When heavy rain comes, however, the small dams are quickly washed away. There are few perennial streams or large water sources which can be used, and irrigation is mainly in the form of opportunistic rainwater harvesting; there is little need for organization and cooperation between water-users.[3]

Rain at times is extremely heavy, and flooding can be a problem in the lower parts of streams. In areas that are known to be prone to flooding (areas which are often valuable land, as they may be able to receive irrigation water), a large wall may be built to protect the fields. These high walls are called *tomota.* The boulders used are so large that these walls can only be constructed by calling together a very large work party; even then, finishing a wall can take months (Photo 1.10).

The Konso keep cattle, goats, sheep and a few hens. Some livestock is kept in the lowlands where a relative or friend will stay and look after it, but most is kept on the lower part of the home compound and zero-grazed. Fodder must therefore be collected daily (often twice a day) and generally from the household's own fields. Trees are kept in the field for this purpose, but weeds and the planted crops which have become over-dense may also be pulled out and taken back for the animals. The manure of the animals is collected and spread on the fields – a process which is vital, because the fields are used year in and year out. In the past, in some

Photo 1.6 *Square basins collecting water after rain*

Photo 1.7 *Enough water has been 'harvested' here to grow bananas by redirecting rainfall that has collected on a road*

Photo 1.8 *An irrigation canal carrying water from a small stream*

Photo 1.9 *The small stream has been dammed to raise the water level*

Photo 1.10 *A* tomota *flood protection wall*

villages (including Buso) the need for fertilizer was so high that they used to use dried human faeces, but this is now very uncommon. It is not that the need for fertilizer has reduced, or that there are alternative fertilizers in use, but that now the use of human faeces is thought 'dirty'. The people of Buso are occasionally ridiculed by others because of these past practices.

Manuring the fields is part of the agricultural cycle. In the dry season (*pona*), the fields are dug over with a special long hoe, and manure and field waste is put deep into the soil. The rest of the year, digging is done with a simple hoe, or with a two-pronged hoe, known as a *bayrra*. The *bayrra* has two iron tips, each about 3cm wide. It allows the ground to be turned, incorporating new manure or in preparation for planting. After the plants have begun to grow, the two-tipped *bayrra* allows the ground around growing plants to be tilled and the area weeded.

A wide variety of crops and trees are grown together on the terraces. The roots of trees help to bind the soil; trees can also attract moisture and provide shade for crops; their leaves can give shelter against rain that could lead to rain-splash erosion; and they can also protect against wind damage. In addition, their falling leaves can provide more waste to be broken down to form humus, and some of them can enrich the soil. The crops themselves are grown together (Photo 1.11). This intercropping not only provides a wide range of crops to serve various uses, but has certain benefits: different crops can exploit different soil niches, and so make more use of the water at different levels in the soil; different crops may provide other crops with other beneficial micro-climates; some fix nutrients. Different crops also benefit from different conditions, and

Photo 1.11 *Intercropped field (maize, sorghum,* qat, *coffee, sunflowers, beans, gourd and other crops are visible)*

cultivating a wide range makes it more likely that the needs of one of the crops planted will get the conditions that they require. In the same way, it is unlikely that all the crops will be attacked severely by pests. In short, intercropping spreads risk. Many of the crops that are preferred are drought- or pest-resistant.

Sorghum is the most common and important cereal crop, followed by maize and millet, and, in the highland areas (over 1800m), wheat and barley.[4] When it is ripening, sorghum heads are tied together so they don't fall or get blown by the wind. Some teff is grown, usually in areas which are flatter and where ox-ploughs can be used. Teff is used to make the preferred food of northern Ethiopians, the pancake-like '*injera*'. It is not eaten by many Konso people (Konso food is made more from sorghum, millet or maize), and is grown largely as a cash crop. It is therefore not common nor grown by many, particularly as it also requires access to an ox-plough. Pulses are another important part of the Konso diet, and they are grown among the cereals. Common varieties include common bean, hyacinth bean, cowpea, chickpea and pigeon pea (a bean that grows on a small tree and is a particular Konso favourite). Mung beans, lentils, horse beans and peas are cultivated, but less commonly. Maize and bean seeds tend to be planted individually. The seeds of millet and sorghum are broadcast. When the millet and sorghum begin to grow, they often need to be thinned. Some thinnings are taken home to feed animals; others are planted in areas where the seedlings are sparse. When all crops are ripening, birdscaring and chasing of other pests (like baboons) is essential.

Trees include *Moringa stenopetala,* the leaves of which are used as the main green vegetable, and are often eaten daily with boiled dough balls made from sorghum and *Terminalia brownii,* known in Konso as *oypata,* whose termite-resistant wood is good for building, coffee trees and *qat* trees. The mildly narcotic *qat* (chat) is rarely used by the Konso themselves, but is sold to *qat* traders who come from Arba Minch to buy it. The Konso also sell the coffee beans directly in the market, or to traders. They themselves make a coffee-like drink from the dried leaves of the coffee tree: they grind the leaves together with garlic, chilli, sunflower and salt, and mix the paste with hot water to make an unusual but effective morning drink.

In addition, sunflower, castor and linseed are grown and used in cooking. Pumpkin, yams, sweet potatoes, potatoes, cassavas and tomatoes are all grown, and make occasional contributions to the diet (potatoes are grown only at higher altitudes). Gourds are grown and used as water or food containers. Cotton is cultivated and spun in the home to make clothes. Tobacco can be grown and used as snuff or smoked, or sold in the market. Generally the fields appear as a riot of different crops with each farmer trying to obtain his or her staple food, combined with a few useful, or tasteful, additions. Fruits can be grown on the few fields that can be irrigated. They include limes, lemons, papayas, bananas and mangoes. They are eaten at home or, as they are relatively rare, they can be readily sold in the village or at the market.

This agriculture meets many of the characteristics of the indigenous technology discussed in the Introduction. Terracing is only one, albeit very important, labour-intensive technique; the Konso employ many techniques together to serve multiple objectives. They harvest water and maximize its infiltration; they conserve soil and improve its fertility; they dispose of excess water and improve drainage on the fields. Many of the techniques, including intercropping and choice of crops, minimize risk; maximizing the use of rain, when it does fall, also insures against dry periods. The choice of a wide variety of crops and the way that they are inter-cropped, also staggers some of the work, particularly harvesting, and prevents, as much as possible, labour bottlenecks. All the materials used in the soil and water conservation are available locally (stones, manure, trash, wood if necessary); manure, wood and trash are renewable resources, and stones are reusable resources; the soil and water conservation can thus be considered sustainable. It represents a combination of different techniques, of the kind commended by Kruger *et al.* (1996) in their inventory of indigenous soil and water techniques in Ethiopia. When I met Kruger, he commented that the Konso case was particularly impressive, as it combines so many different techniques to maximize yield whilst also minimizing risk; it is difficult to see how it could be improved.

Konso is not only an impressive case of indigenous agriculture and indigenous soil and water conservation; it also represents a form of intensive agriculture. In the following section I review further what this means, why it is important, what can be said about how it came about, and some of the other characteristics of this kind of agriculture.

Intensive agriculture

Intensive agriculture requires more skills, energy or capital to be invested into the land, usually to gain more output. If soil fertility is declining, intensification will be needed just to maintain the previous output level. Hakansson describes three kinds of investments which facilitate it: labour intensification; capital intensification; and change in technology (Hakansson, 1989: 14). 'Labour intensification occurs where the effort and time spent in cultivation is sustained on a higher level than previously, resulting in higher yields per unit of land' (ibid.: 14). Netting refers to this when he writes that intensification 'has a cost… the cost is often human energy, the product of sinew and sweat' (Netting, 1993: 103). 'Capital intensification' is a related concept, in that the production of 'capital', which facilitates increased production, frequently requires the investment of more labour. The form that this takes is Blaikie and Brookfield's 'landesque capital': a 'class of works, including stone walls, terraces and such improvements as field drains, water meadows, irrigation systems and regional drainage and reclamation systems' (Blaikie and Brookfield, 1987: 9). These structures are created with the intention of maintaining or improving land capability. Some are soil and water conservation structures; some allow an improved yield. The construction of landesque capital does not necessarily provide immediate returns; the goal is obtaining returns in the future. Landesque capital 'refers to any investment in land with an anticipated life well beyond that of the present crop, or crop cycle' (ibid.: 9). The final form of change which produces intensification is 'technological intensification'. This 'includes new varieties of plants and the establishment of radically altered ecosystems geared towards agricultural production such as crop diversity and shorter fallow periods. This category might also include new implements and pest control' (Hakansson, 1989: 14).

There are different forms and degrees of intensive agriculture depending on the combination of the above. Hakansson (1989), for example, describes how if labour-saving devices are available and are adopted by farmers, a more productive system may be established without an increase in labour demands. Widgren (2004) explores the way in which some intensification is accompanied by soil and water conservation structures, many of which are forms of landesque capital such as terraces, but others are not. Konso is an example of intensive agriculture that combines many of the criteria above. It is labour-intensive; it constructs large amounts of landesque capital in the form of terraces, drainage structures, irrigation furrows, and flood protection walls. Technologically, the Konso grow many different species of the same crop, and combine these and other crops together to minimize risk and to improve their yield. The fields are cultivated permanently, and manure is applied to attempt to prevent soil exhaustion. Cultivation on the terraces is restricted to hand tools.

If it involves landesque capital formation, then intensification constitutes a transformation of the environment and the landscape. There are constraints to this transformation, depending on climate and soil properties, but these processes enable people to go beyond some of these environmental constraints. In dry regions they can make the most of the rain that is available, and the soil 'capability' can be improved through techniques such as those already described (Blaikie and Brookfield, 1987). Landesque capital improves the 'carrying capacity' of an environment, as intensification usually allows higher populations to be supported on the same land. Table 1.1 (reproduced in slightly adapted form from Netting, 1993) shows different degrees of intensification of agriculture and the population density each supports. It also shows the frequency of land use, and the extent to which land can be used permanently or is left to fallow.

The significance of these cases of intensive agriculture has already been discussed in brief. In Africa, there are fears that the number of people is outstripping food supplies, and that pressure on resources is leading to

Table 1.1 *Generalized land-use types, fallowing, cropping frequency, and population density (adapted from Netting, 1993: 264)*

	Land-use type	Period sown	Period fallowed	Fallow vegetation	Population density (per/km^2)
Less	Forest fallow	1–2 yrs	15–25 yrs	High forest	0–14
Intensive	Bush fallow	2–8 yrs	8–10 yrs	Low forest	4–16
	Short fallow	2 yrs	1–2 yrs	Grass	16–64
More	Annual cropping	1 crop/yr	Few months	Grass or none	64–256
Intensive	Multicropping	2 crops/yr	No fallow	None	256+

land degradation. Scholars have called for the most alarmist claims about population pressure, soil erosion and desertification to be reassessed, but concerns about population pressure and soil erosion still remain. In 2003, a United Nations Development Programme website, for example, proclaimed that

> Perhaps more than any other population group in the world, Africa's 650 million people are immediately and directly affected by environmental degradation, food scarcity, and population growth.[5]

Kruger's work on Ethiopian soil and water conservation cites statistics 'that 1.5 billion tonnes of soil are eroded every year in Ethiopia...' and that 'between 1985 and 2010 the rate of land degradation will cost 15.3 billion Ethiopian *birr*' (Kruger *et al.* 1996, 171). Many development programmes are oriented to improving soil and water conservation and to population control.

Putting aside for the moment the assessment of the way in which environmental degradation is defined and measured, the intensive agriculture of Konso, and other cases like it, can be viewed as success stories where people have invested their knowledge, skill and labour into their environment in order to support a relatively large population. The Konso experience is similar to the much-cited example of Machakos in Kenya, which in the 1930s was considered an environmental disaster area as a result of population growth and unregulated use. By the 1990s the situation had been turned around through the construction of terraces and the use of other intensive farming practices (Tiffen *et al.* 1994). In Machakos, the achievements in soil and water conservation have been brought about partly by external encouragement, and partly by intervention, by colonial agricultural officers and later by development workers (although the achievements appear to be most significant when taken up, adapted and implemented by small-holder farmers themselves). Cases of indigenous intensive agriculture, such as that of Konso, are exciting for contemporary scholars and development workers because they also buck the Malthusian trend: they show that 'more people' do not necessarily equate with 'more erosion' or more people going hungry. In these cases, the farmers have implemented the intensification processes themselves, long before colonial officers came, concerned to encourage people to build terraces (Anderson, 1984). The Konso, and other extant cases of indigenous intensive agriculture, have continued to support people from their agriculture for many years, whereas for one reason or another, many of the other forms of intensive agriculture have fallen into disuse.

Why intensify in Konso?

Much attention and research has been directed to examining how and why intensive agriculture emerged in particular places and how it has been sustained over time. It is not necessary to rehearse all the arguments in detail, and I will draw out only the main points which are relevant to understanding Konso. More thorough treatment of these debates can be found in Widgren and Sutton (2004), and Netting (1993).

Initial ideas that cases of intensive agriculture corresponded to particular physical environments have now been dismissed, as a great deal of variability has been found both within and between the physical environments in which they are found. Climate, soils, altitude, and the location of permanent water sources do not correlate with whether or not intensive agriculture develops in different places. Physical conditions may provide parameters which make intensive agriculture possible, but little more (Widgren, 2004; Netting, 1993). Theorists of intensive agriculture have thus shifted their attention to population and natural resource dynamics, particularly following the influence of Boserup's 1965 work, *The Conditions of Agricultural Growth.* Boserup demonstrated that population pressure does not necessarily lead to environmental degradation; it can also lead to increased investment of labour and to innovation, hence intensification. For Boserup, the necessity of feeding more mouths is the mother of invention and the impetus for harder work.

Netting supports and extends Boserup's approach. He contends that intensification results from the 'spur of scarcity' (Netting, 1993: 263). He emphasizes that population pressure can result not only from population growth, but also from constraints of movement, exclusion from resources, or an influx of people:

> The practices and technologies that the smallholder uses to intensify production from limited resources are means of coping with constraints on space and time. The shifting cultivator who lacks land because of encountering a geographical barrier like a range of mountains, the increased competition of a growing population of other farmers, or the removal of land by a warring enemy or plantation owner is faced with the problem of occupying and using the remaining land more continuously. Smaller available space means some combination of producing more crops per unit area and using the area for a longer period in time (Netting, 1993: 102).

The scenario that the 'removal of land by a warring enemy', leads to population pressure and intensification, is an example of the 'siege hypothesis', one of the most popularly held explanations for intensification in Africa. Political unrest, particularly inter-ethnic conflict, prevents free movement and use of natural resources; land in safe areas provides a refuge for people from surrounding insecure regions, creating population pressure. Population pressure may also result from a particular area becoming a refuge from other difficulties: highland areas may provide a refuge from insect-borne diseases, more prevalent in lowland regions; once intensification practices have begun, areas of intensive agriculture can also provide a refuge from surrounding areas in times of drought.

Studies of intensification that have drawn on Boserup's thesis also emphasise the role of the market, in addition to population pressure, a force leading to intensification (Netting, 1993; Tiffen, Mortimore and Gichuki, 1996). The market exerts push and pull factors: a non-agricultural area or food-deficit region makes a demand on production; exchange and

profit incentives may contribute to the intensification of production elsewhere (Netting, 1993). External pressures such as increased taxation or calls for tribute payments can also create the demand which leads to intensification (ibid.: 1993).

In Konso there is a high population density. The travellers' accounts at the beginning of this chapter have shown that in the past the lowlands surrounding the Konso highlands were inhospitable: insect-borne diseases were rife and wild animals were abundant and often dangerous. The Konso have frequently been in violent conflict with their neighbours, particularly the pastoralist groups of the Boran, Tsemai, and Guji (Map 1.2). The defensive walls in most of Konso are evidence that in this area there may be some validity to the much-maligned siege hypothesis.

The oral histories of the Konso people concerning their origins are diverse: some claim that they are autochthonous, that the first Konso person was born from a gourd in Konso and then gave birth to the rest of the Konso people. But other origin myths claim that they came from different directions: some from Boran to the south-east (where one myth states the first Konso person was born from a branch of a tree, together with the first Boran person). Others claim they came from a mountain, 'Ala', in Gawwada (to the north-west of Konso; see Map 1.2). One origin myth even claims that a founding lineage came from a place called Kontari, so far north that it is now assumed to be Gondar.[6] This diversity in narratives of origin seems compatible with the idea of Konso as absorbing people from all the surrounding areas, who were willing to settle in this hilly place and cultivate. In Konso, it is also possible for a person to be adopted into the community: such a person is known as *orrhayta*, and over time they may become an accepted member of the clan, lineage and household. As the agriculture is labour-intensive, an extra pair of hands is always useful, and it is likely that people from outside were welcomed. It is possible to infer that people may have settled in Konso from the regions all around. This settlement may have expanded in times of insecurity, drought or food shortage, when Konso provided a refuge, and this expansion provided both the labour and the demand for intensification.

In the Kenyan Machakos case, proximity to the market of Nairobi proved to be a strong spur for intensification. In Konso, although there are urban markets in the region today, the Konso engagement with these markets cannot be compared to the scale of Machakos' engagement with Nairobi. In the past, when intensification developed, these activities would have been minimal. But Konso is surrounded by pastoralist peoples (Boran, Guji, Tsemai), and there have been extensive exchanges of milk products and livestock in return for grain.

A final factor that might be said to be conducive to intensification is the environment in Konso itself. The Konso highlands are volcanic, which means that some of the soils are likely to be, or to have been, fertile. Moreover, the land in places is very stony, and these stones must be collected up in order to cultivate. Donkin's work on terracing in Central and Southern Mexico makes clear that terracing requires a great deal of hard work, skill and imagination, but also that certain physical environments and processes might suggest it: 'Contour terracing… may have been prompted by the accumulation of soil behind rock outcrops, fallen trees, the walls of dwellings and, in the Andes, animal pounds. The next step could have been the deliberate alignment of loose stones where it was necessary or customary to remove these in the course of cultivation' (Donkin, 1979: 131). It likely to be the combination of these different factors which contributed to intensification in Konso.

Social characteristics of intensive agriculturalists

The above accounts of intensification explore the historical and ecological processes leading to intensification, somewhat in the abstract. As described in the introduction, the knowledge, skills and techniques produce the indigenous agriculture (in this case intensive indigenous agriculture) in an institutional context. This institutional context refers to the social processes which shape the use of, and access to, the key resources of land and labour. The very word 'labour' has the effect of reducing the act of work on the landscape to an act that is part of the population-resources equation. If the word 'work' is used instead, it is more likely to encompass an appreciation that agricultural 'work' is socially organized and culturally valued, as is any other kind of work. People in Konso may need to till the ground to make a living, but the way in which their physical effort is applied depends on the way in which they perceive and value their work. Sutton (1989) also argues that cases of intensive agriculture are not just complex combinations of different technologies, but are the result of the application of different survival strategies (in which there may be some choice) and feats of social organization. As such they must be understood as much socially as physically. The changing social organization and value attached to land and labour are the subject of the following chapter, but it will be helpful here to examine how far Konso shows evidence of the characteristics that Netting identified as being present among most communities that practise intensive agriculture (1993: 2–3). These refer to the social characteristics of an 'ideal type' he describes as 'smallholders'. Smallholder livelihoods are characterized as 1) household-based; 2) they are diversified, largely subsistence-based, but with some market participation; 3) they have well-defined and long-term tenure rights; and 4) there is often some inequality between households. In reviewing these characteristics and exploring how they relate to Konso, it is possible to introduce more details about Konso, and to set the context in which to examine the changing social organization of land and labour in the next chapter.

Netting's characteristics and their relevance to Konso

Firstly, among intensive agriculturalists, production is organized, according to Netting, around the social unit of the household. Here the influence of Chayanov's economics is evident: the household is an effective and resilient unit of production (Chayanov, 1966). It has the capacity to 'self-exploit' at times of need by digging deeper and working harder, for instance when the development cycle of a family is such that the number of consumers is higher than the number of workers. This capacity is particularly useful in intensive agriculture, which requires large amounts of labour.

In Konso the household is a most important social unit. The Konso are patrilocal, and inheritance rights to residence are based on the principle of primogeniture: the eldest son will always live in his natal homestead, ultimately inheriting it, and looking after his father and mother in their old age. He also inherits and keeps the name of the founding father of his branch of the lineage. In contrast, on marrying, younger brothers must leave and build their own home, a costly and time-consuming process. Younger brothers are known only by their given name at birth and by their father's first name, not by the name of the founding father of the lineage. Thus, each younger brother is himself the founding father of a new branch of a lineage. The difference between the rights, responsibilities and values attached to elder and younger brothers is wide and taken for granted. It is also institutionalized through naming: *all* second sons (*kussitas*) in Konso are named 'Kussi'; they are not given another name, and this serves to make a clear distinction between the lot and life of the first-born and that of the rest from birth.

Among polygynous households[7] each wife should have her own home and fields to cultivate. The first wife is always considered senior: she is referred to as the wife and mother of the family (*tola*), whereas subsequent wives are referred to as *lamiteta*. The households of second or third wives are considered to be headed by the woman, not by her husband – the husband is 'head' of *tola*, the first wife's household. The consequence of this delineation is real enough: it is usual for the husband to pay tax for his first household, his '*tola*', whereas the second wife pays tax for her own homestead. In the same way as second sons of first wives are called 'Kussi', the first children of second wives are called 'Lamita'. Thus production is very much household-based, as would be expected from Netting's model. Households may be linked through a shared husband and father who is unlikely, however, to contribute equally to different households.

The second of Netting's points refers to the diversified nature of livelihoods, their subsistence orientation, but also to their participation in the market. As we have seen, the Konso cultivate as diverse a range of crops as could be imagined. Few households sell subsistence crops, but other crops are often traded, usually through the market place. The main markets are in Karate, in Dokato, and Fasha, taking place approximately twice a week. The centrality of market days in Konso life is reflected in the way the names of these market places are given to the names of the days of the week on which they are held. For example, Monday is Umbocko (in Dokato), Tuesday is Umbocko Parayta (the day after Umbocko), Wednesday is Gomosa, Thursday is Bekawle (in Karate), Friday is Bekawle Parayta (the day after Bekawle), Saturday is Idigle (in the east of Konso), Sunday is Sambada (different in that this is probably derived from 'Sabbath', showing the influence of the Christian religion in the region). Thus the passing of time is associated with the gathering of people at these different market places. Going to market is a great opportunity to meet friends and relatives from other villages, as well as to buy and sell. Markets are held from late morning through to the afternoons, and much Konso beer (*chagga*) is drunk in the houses that surround the market.

As well as diversification within agriculture, there is a diversification of livelihoods more broadly. There are people who live from craft working: from making pots; from tanning hides and working leather; from weaving cotton; and from working iron. Included in this group are also butchers, who buy livestock and slaughter animals in the market. This group of people is known as *xawuda*, a term that is used to refer to people who are craftworkers and traders (Hallpike, 1968). In contrast, farmers are referred to as *etenta*. *Etenta* and *xawuda* represent two occupational social groups, and in the past marriage between them was socially prohibited. The two groups have strong social identities; they live together in the same neighbourhood and they co-operate together, but there are differences in their everyday activities and lifestyles; there is sometimes antipathy between the two groups that varies from gentle teasing to outright hostility. The agriculturalists see themselves as hardworking, honestly tilling the soil and harvesting and storing their means to live. They view the *xawuda* as living a hand-to-mouth existence, forced to rely on trade for subsistence, a trade that sometimes involves smelly and unpleasant work, for example the tanning of hides, or the selling of blood in the market place (see Watson and Lakew, 2003; Freeman and Pankhurst, 2003). A *xawuda* blessing, that was told to me by a *xawuda* elder, demonstrates how they collude in this representation of themselves as quick, crafty, and profit-oriented, in contrast to the slow farmers:

Xawudaitaayo a xanta minna sigado	May our *xawuda* fly like a bee
Kannatadi kannata qara qatu	May our deals be more and more[8]
Halbadi a korita q'oto	So that one half will be made into a full cloth
Kera afurrotoni pato	A thief will sell the bead for a good price

The farmers' rudeness to *xawuda* can be seen in a well-known weeding song in which one line goes 'the ancient *xawuda* are eating the inside of a skin and belching terrible smells'. In retaliation, the *xawuda* refer to the farmers more mildly as 'weed pullers' (Watson and Lakew, 2003).

In recent years, farmers have been diversifying their livelihoods significantly. Labour migration is common: men go to work in nearby towns, particularly in the border town of Moyale where they carry goods for traders across the border. In these ways, the Konso agriculture meets Netting's second criterion of being 'diversified, largely subsistence based, with some market participation' (1993: 2–3).

The third of Netting's characteristics relates to land tenure. According to Netting, the investment of labour on to land that is cultivated permanently leads to long-term and heritable land rights: 'where land is a scarce good that can be made to yield continuously and reliably over the long term by intensive methods, rights approximating those of private property will develop' (Netting, 1993: 158). Landesque capital, being highly visible in the landscape, constitutes a long-term claim to land, effectively privatizing the commons. It is necessary to be cautious in the application of terms like 'private property' and 'ownership', to avoid implying a set of relations to land that exist in one society (usually Western) to another (Bruce, 1990). In Konso the term 'ownership' can be applied to indicate 'prime right of property over the land enabling a relatively high degree of exclusivity of use, the right to sell or otherwise alienate and the discretion to pass it on to heirs' (Ward and Kingdon, 1995: 9). Such rights have developed with the intensive agriculture and they represent the 'customary land tenure' practices in Konso.

According to customary rights, original ownership of land is established by cutting or burning previously uncultivated land. Subsequent ownership of land is by inheritance or buying and, as stated above, 'ownership' means the right to use, to sell, to lend or to pass on to heirs. Only men own land; women do not inherit land. It is possible that if a woman bought or cleared land herself she could own it, but such claims are unusual. Netting (1993) and Linares (1992) have shown that when pressure on land becomes extreme, and investment of labour into the land is high, it is not only the individualization of land holdings that tends to increase: inheritance also tends to be stratified and is biased towards one descendant. This is the case in Konso: there is no full primogeniture, but inheritance is heavily weighted to the eldest son, the embodiment of male line from the founding ancestor. The example of one Buso family with four sons can be used to illustrate this: the first son inherited 11 fields from his father; the second son inherited 7 fields; the third son inherited 3 fields; the fourth son inherited 3 fields also, but these fields were mostly 'stony and infertile'. The fourth son found it impossible to make a living from his poor fields. He borrowed money and managed to purchase some fields in the lowland area of Fuchucha, following the fall of the *Derg* (the 1974–1991 government). The yield in this area is good, but he and his family have suffered from recurrent malaria, prevalent in the lowlands.

These customary tenure rights exist within a broader national context in which all land has been nationalized, and made the common property of the Ethiopian people. In 1975, following the revolution in 1974, 'all private ownership of land by individuals is prohibited; so too, the transfer of land by sale, lease, mortgage or similar means' (Dessalegn Rahmato, 1984: 37). Despite the change of government in 1991, the situation remains very much the same. Thus two 'bundles of rights' (Bruce, 1990: 1) have the potential to provide rights to land in Konso: firstly the customary rights provide individualized rights to land, developed from the investment of labour into the land and its intensive use; and secondly the statutory rights provide only usufructory rights. The way these contrasting and potentially conflicting regimes of rights are worked out is explored later in this book. The rights to land that are discussed in Chapters Two, Three and Four refer mainly to customary rights to land. This emphasis on customary rights is not an attempt to deny the impact of the land reform on Konso, but recognizes that the customary land tenure practices are still strong. These chapters focus on unpicking the relationship between these landholdings, labour practices, social relations, belief systems, and the indigenous intensive agriculture that represent the orthodoxy as discussed in the introduction.

The final characteristic that Netting discusses as being present among intensive agriculturalists is inequality:

> the existence of separate household enterprises, with a measure of autonomy and self-determination, in a larger economy with institutionalised property rights and market exchange, presents the likelihood of economic inequality, both among households in the community at any point in time and in the changing status of a single household at different times in its development cycle (Netting, 1993: 2).

Although few communities in Africa fulfil the stereotype of being egalitarian, inequality is particularly well institutionalised in Konso. It has been shown already that inheritance is unequal between brothers; the landed farmers and the landless craftworkers could be seen as another example of inequality. Gender inequality is evident in land ownership, participation in decision-making fora, and in work patterns and loads. Further axes of inequality exist and emerge as a central part of discussion in the subsequent chapters.

Having discussed the indigenous knowledge and technologies that make up the Konso agriculture, and introduced some of the social institutions that accompany the intensive agriculture, the following chapters look in more detail at the social and cultural institutional dimensions of this agriculture. However, it is now clear that the agriculture is impressive, flexible and innovative; but it should also be added that the agriculture is far from fail-safe. Food shortages have been common in the past, and they are common today:

> Despite untiring efforts to minimize the risk, famine, hunger and starvation are characteristic of the majority of the peasants. Since the beginning of this century [the twentieth], the people of Konso have borne the brunt of six major droughts all of which have resulted in fatal famines... In general, since 1950, 'normal' crop years and years of meagre crop production have occurred almost alternately. Low agricultural production is characteristic of the majority of peasants (Messeret Lejebo, 1990: 35).

One elder in the community explained to me that only in some years would there be enough rain for a good harvest, and that food had to be stored from these in order to cope with lean years. His observations were also that the ratio of good to bad years had worsened recently and was exacerbating food shortages:

> In the past, if there were a good yield in the field it would last for even two or three years. In ten years there must be at least three good years of satisfactory rain. If the rain is less than this there will be drought. For seven years now there has been poor rain.[9]

The pattern of uneven harvests has continued in the twenty-first century. In 2000 and 2003 there were severe food shortages, and food distributions have become routine. In 2004, 2005, 2006 and 2007, the harvests were good and grain filled the granaries. In 2008 drought has hit the wider region of Ethiopia, northern Kenya, and Somalia, and Konso has been experiencing food shortages again. The bad years demonstrate that caution is required when celebrating the intensive agriculture of Konso. In a region where food shortages are widespread, however, it does not mean that Konso agriculture should be viewed overall as a failing system.

Emic views on Konso livelihoods and landscape

I described earlier some travellers' responses to the landscape and livelihoods they encountered in Konso. It is interesting to contrast these with the views of the Konso people themselves on their own landscape. This is not easy, as they do not wax lyrical about the landscape as a thing in itself to be celebrated; nor is there such a thing as a monolithic set of 'Konso ideas' about the landscape. But some particular attitudes to Konso as a place are discernible, and also some expressions can be used to give a flavour of different ideas and attitudes.

Firstly, the highlands and lowlands have very different connotations. For example, the word Konso (or 'Xonso' as it is pronounced in Konso[10]) is used to refer to the highland area itself, derived from Xonso hill (near Dokato). Konso is the land that is worked and settled. It is contrasted with the hot, lowland areas that surround Konso that are collectively known as '*Gommaide*'. Some men place beehives in trees here, as the plentiful bushes provide flowers from which to make honey; some people raise livestock, and it is also a place for hunting. Fighting has also taken place in Gommaide between Konso and their ethnic neighbours in the past. As already mentioned, it is also a place of disease, and of attacks by wild animals. Overall, Gommaide is seen as a dangerous and hot place, compared to the ordered, controlled and cooler environment of the highlands. In the highlands themselves, there are pockets of forest, much reduced over the last fifty years, but nonetheless still present. These are seen as peaceful places in which there is thought to be *xaixaita* (an onomatopoeic word meaning wind or breeze) that is good, cooling and restful.

Secondly, the Konso have particular perceptions of their own livelihoods. If Konso are asked to tell about their own lives, the accounts have some similarities to those of the travellers. Again, they celebrate their hard work:

> Konso people like work very much because it is all that we can do. Really we like work: our women get up in the middle of the night to start work, to grind or to collect water. We men get up in the middle of the night to make beehives, to go to the lowlands, or to collect forage for the animals. The best work for men is farming, working the land and producing the crops.[11]

A woman's account does not stress so much the liking for work, but the amount that she has to do:

> At the first cry of the cockerel we awake and grind the grain, then we make food, then we go for water, then we make coffee, then we give coffee to those in the house, then we go for fodder for the animals. Then, when there is rain, we go to the fields.[12]

Gommaide is different from Gummayde, which is an area to the north-east of Konso where many Konso have settled in the last thirty years or so (see Map 7.1, Chapter Seven, p. 202). Those who have gone to live in Gummayde, or to the lowland areas surrounding the Konso highlands, agree that life there is not always as hard: the land is less 'stony', and the cultivation is less intensive. Often they need to clear new land, but once this has been achieved, if the rains are good, the land can give high yields. If the rains are not good, then a harvest in the lowlands may be completely lost. The multiple cropping and soil and water conservation in the Konso highlands means that, though harvests may be low, they are not often lost altogether. There are other advantages to living on the hills of the Konso heartlands. One young man who had moved to Gummayde, and who came to Konso only every few months, told me how he missed his village. Konso is a crowded and intensive social centre where friends and relatives are in close contact. In contrast the frontier Konso areas provide a livelihood, but life there feels comparatively isolated.

Notes

1 See Fairhead and Leach, 1996, for a discussion of the methodological implications of using travellers' reports for constructing histories.

2 '*Le Konso proprement dit est un curieux petit pays récemment conquis par les Abyssins, et dont l'étendue, dans la partie dont je veux vous entretenir, ne dépasse pas 20 kilomètres. Cette partie, montagneuse, est peuplée par des individus complètement noirs, dont le langage et les usages diffèrent de ceux des pays environnants. Ils sont remarquables par leur amour du travail. Leurs champs de café, de coton, de maïs, disposés en gradins sur le flanc des collines, sont soignés d'une manière irréprochable. Leurs maisons, symétriquement construites sur des pilotis qui les préservent de l'humidité, sont parfaitement propres*' (Darragon, 1898: 138–9).

3 There is some more extensive irrigation in the lowlands, on the Yande furrow below Karate town, and in the area of Konso that borders with the Gawwada and the Tsemai.

4 Some of these crops are of relatively recent arrival in Africa, demonstrating that Konso agriculture has been adapting to new opportunities and possibilities for a long time.

5 http://www.sdnp.undp.org/sdncmr/adi/pop.htm; accessed 6 August 2003.

6 This is the origin of the Kalla *poqalla* family. Their origin in Gondar seems unlikely (it is more probably a subsequent explanation derived from the similarity between the names Kontari and Gondar, see Hallpike, 1972).

7 In Buso 10% of a sample of 145 households were polygynous.

8 Literally, 'may our hands be clapped together again and again': hands are clapped to signify a deal.

9 Kes Beyene, 2002.

10 In *Alfa Xonso*, the 'X' sounds somewhere between an English 'K' and 'H'.

11 Samuel (Konso man), 22 July 1995.

12 Galiso (Buso woman), 2002.

Two

Social Life of Agriculture

The customary agricultural practices and institutions that are described in this Part of the book are those practices and power relations whose legitimacy is grounded in the claims to tradition, to the ways things have always been and should be. Using Bourdieu's terms, this represents the orthodoxy in Konso. The title of this specific chapter is borrowed and developed from Appadurai's edited book, *The Social Life of Things* (1986). Others have taken up his approach before now, for example Nyerges (1997) uses his ideas to talk about 'the social life of resources', and Longley (2001) uses his ideas to talk about 'the social life of seeds'. For these authors, Appadurai's approach to things is useful because it sees them not as inert, but as objects with 'careers', 'life histories' and 'biographies', which are 'fully part of social life' (Nyerges, 1997: 12). Thus the environment is presented not merely as a backdrop to social life, but as deeply inter-connected with, and part of, that social life. This approach has therefore much in common with the approach to agriculture as a social process discussed in the introduction. In the next chapter, the approach is taken further to show that not only are the practices and institutions part of social life, they are also part of ritual life. The title the 'Ritual Life of Agriculture' is used to show that these social processes are also cultural processes too, replete with ritual actions, relationships and meanings. In this chapter, the title the 'Social Life of Agriculture' is used to highlight the way in which agriculture is shaped by the constraints, desires, arrangements, performances and relations that are social.

Work in the household

As described in the last chapter, the household is one of the main units through which work is organized and the intensive agriculture is produced. Within the household, the work is structured by socially constructed and

gendered power relations. In Konso, certain tasks can be performed by either men or women, and sometimes men and women work together. This is different from many African societies where there is a strong differentiation between men and women's work, and men and women's spheres of activity. Here, the shared tasks include digging the soil in preparation for sowing, sowing itself, weeding, harvesting, and scaring birds. Both men and women collect fodder and carry it back to the homestead to feed the animals that are tethered there, but a distinction is made in the way men and women carry out this work: women carry their loads on their backs, while men carry them on their shoulders or their heads. If a man carried anything on his back then it would be said that he would be behaving 'like a woman'.

It is men who construct the terraces and other soil and water conservation structures, although women may contribute by carrying soil and stones to help in the construction and ground levelling process. Men are responsible for fencing and for maintaining the homestead, public buildings, and pathways. Fences are mainly around the homesteads and along the sides of roads and pathways. Cattle are not allowed on the terraces as they would cause damage, and the borders of adjoining fields are marked by prominent stones, or particular plant species, and do not need to be fenced. Beekeeping and hunting are also male activities.

Women play a large role in the everyday tasks associated with the agriculture, but they are particularly associated with the collection of manure and spreading it on the fields in order to ensure soil fertility. Sometimes men also perform this work, but usually it is carried out by women. Women fetch water and firewood; they grind grain and prepare food. They also carry out the majority of the childcare, although it is not unusual to see a man caring for a child. In addition, women are exclusively responsible for brewing. This is a process that demands a large amount of time, labour, water and firewood, and it is an integral part of the agriculture. *Chagga* (beer) is essential if a household is calling a work party, in order to sustain the workers and reward them after they have finished.

Chagga is made from sorghum, millet or maize. It varies in alcoholic content from mild to quite strong. It is often drunk with the addition of hot water, so it is warming physically as well as psychologically and it fills the stomach. In order to prepare *chagga*, a large amount of grain must be ground and fermented. A distinction is made between *chagga* that is brewed for the family (*tola*) and *chagga* that is brewed for sale. In *chagga* for the family, grain from the fields of the family may be used, but in *chagga* for sale, the woman is expected to purchase grain from the market. Thus brewing for sale is a commercial activity from start to finish, and most women brew and sell *chagga* from time to time. Some women rely on this activity for a living (especially female-headed households). A strong clear liquor can also be distilled from *chagga* known as *arake*. This involves more labour and fuel to heat the beer for distillation, but it produces a higher income.[1] Much of the income of men and women is pooled in the

Photo 2.1 *Mixing ground sorghum, maize and millet to make* chagga

Photo 2.2 *Boiling the grain balls in the process of making* chagga

Photo 2.3 *Girl carrying dry sorghum stalks*

Photo 2.4 *Koyitti and Tabach (Xalale and Kahana's daughters) grinding by hand*

household, but when women bring in income they may have an increased ability to retain control over some of it, and/or have a stronger say in how it is used.

The work of children in Konso is also highly valued: they work on the fields with their parents from an early age, helping with sowing, weeding, thinning the crops, and harvesting. From around the age of five they are learning about different plants and techniques. Young children, male and female, play an important role in bird-scaring, and just before the harvest they may spend the majority of their time in the fields, 'playing' with children in neighbouring fields, honing their skill at hitting birds with a stone from a sling-shot. Young boys and girls also care for their younger siblings, and young girls play a major part in collecting water from a very young age. In this way, to use Donham's terms, work is both gendered and 'engendering' (Donham, 1990): it is through these tasks that a child learns how to be a Konso person, and through the gendered roles they learn about how to be a man or a woman. They develop different skills and come to know what is expected of them.

A large amount of labour falls to women, and women's labour is highly valued. I came across many cases of eldest sons (who will always remain in their fathers' homesteads) at an age as young as 12 or 14, who were being pressurized by their parents to marry, so that they would bring another pair of hands into the household. Women are therefore often busy. If they want to meet each other, they usually go from homestead to homestead, where they help each other with work: preparing *chagga*, or Konso coffee (*xolla*), or spinning cotton, or many other kinds of work, whereas men who want to meet and chat often go and sit in public meeting places in the village, known as *mora*.

Thus the household is a most important unit in the construction and maintenance of the agriculture and the landscape. Men, women and children work together on the fields, and this field work is part of a broader sphere of work in which their subsistence is maintained. This is Netting's 'smallholder', Chayanov's 'peasant farm', or the 'domestic mode of production' described by Sahlins (1974). The household exists as a productive, consumptive and social unit. The members work together, eat together, and together they keep the animals which provide manure. The individuality of the household is marked in other ways. In the densely packed villages, each household is bounded from the next by a high fence, to stop prying eyes; conversations often take place in the homestead in a whisper, to prevent neighbours from being aware of the family's business. A strong distinction is made between the *tiga* (house) of the family, and *tigalayyo* – another's house. Someone who visits another too often, and eats their food, is frowned upon for spending too much time 'in *tigalayyo*'. But, as elsewhere, identities in Konso are multiple. Although the family household is central, socially and productively, each member of each household is also part of a wider network of connections, obligations and activities, which cut across the household. Each household member is also a member of a clan, a village, and a neighbourhood. Women in particular are connected to the family, neighbourhood, or village from which they married. Men and women both have their own networks of friends and relations which provide the basis of a rich social life, and, in various ways, are also important institutions in the production of the agriculture and the landscape.

Networks of cooperation: labour between households

Early in my time in Konso, I was woken by a horn blowing in the distance. I went to the road and saw people emerging from a village, each with a *bayrra*, the two-pronged hoe, swung over one shoulder and carrying a small gourd filled with water or *chagga*. These people varied in age: they were mostly young men, but there were some young women, and one or two older men. They walked together hurriedly down to the lowlands, almost running, and occasionally breaking into song. It resonated with Hallpike's description of Konso people going to the fields as other people go into battle. I followed them down to the lowlands, and watched as they moved energetically up and down the field in a line, digging the soil over ready for the next rain, keeping in rhythm with each other by singing as they went along (see Photo 2.5).

At one side of the field there were two women, who tended a fire and heated water. The women had brought water, *chagga* and beans. After a couple of hours' work, and when the sun was very hot, we sat in the shade of a tree, and everyone ate, drank and rested for a while. There was much laughing and joking. It was my first introduction to a working party, an impressive and hardworking group of people from Baddingaldo village. The word 'party' is used usually to infer group work rather like a military posse, but in the Konso case the word 'party' is not inappropriate: through working together the work is a social event as much as a task, and makes the work enjoyable. People throw themselves into the work and they sing while they do so. In big work parties they also occasionally stop the work to dance; the work is literally performed and enjoyed as much as it is executed.

Much of the work required for the intensive agriculture is performed and received through participation in work groups such as this, that cut across individual households. They exist as networks of co-operation. They are the forms of social capital and forms of indigenous social institution that make the agriculture and the landscape possible. They embody the qualities of trust, reciprocity and mutual obligation that development practitioners have found so valuable and yet so difficult to construct where they are not already present.

Participation and membership of these networks of trust and cooperation depend on social identity (Berry, 1989a, 1989b). Some forms of

Photo 2.5 *At a particular moment in the singing, all in the work group raise their two-pronged hoe* (bayrra) *and call out*

Photo 2.6 *Lunch of beans and* chagga *after the morning's digging*

identity are more ascribed than others: clan and neighbourhood identities are fairly fixed and given; membership of networks of friends or more formalized work parties may depend more on a person's skill and performance. Social skills and networks may be as important as agricultural skills and ability in determining whether a person is invited to join a group. In subsequent chapters, religious identification is also seen to influence membership of work groups.

Neighbourhood

The neighbourhood (*kanta*) is one of the most important institutions defining identity and structuring co-operation between households. Buso village has five *kanta*[2], and the spatial boundaries and social membership of each neighbourhood are well known and fixed. It is thought that a man who is born in one neighbourhood should not go and live in another.[3] If he does so, misfortunes are thought likely to befall him: it is thought he or his family will fall sick, become mad, or are unlikely to be able to have children. The importance of the neighbourhood as an institution of co-operation is encapsulated in the fact that if a person dies then it is the neighbours' responsibility to bury that person. There are also other neighbourhood responsibilities: if a person is sick, the neighbours carry him or her on a stretcher on their shoulders to the clinic in town. If the paths in the area become overgrown, or if the *dina* – the area of trees that surrounds the village and which is used as a waste ground and a toilet – becomes too dirty, then members of the neighbourhood are called together to clear the paths, or to clear or burn the rubbish. The neighbourhood is also a modern administrative unit. When the villages have to pay taxes they are collected from neighbourhood to neighbourhood. If there is a food distribution, then food-for-work and the food distribution is organized on a neighbourhood by neighbourhood basis. In 2000 and 2001, national and regional elections took place in the village, and these too were organized along neighbourhood lines.

The neighbourhood gives a spatially defined identity, but membership of the neighbourhood also depends on participation in its activities and responsibilities. Without the social participation, the spatial identification is meaningless. If a person does not cooperate with his or her neighbourhood, then initially he or she will be fined and any income is used for the public concerns of the neighbourhood. If a fine has no impact on the behaviour of the individual, then the neighbourhood will refuse all co-operation with him or her; in *Afa Xonso*, they say they will 'shut the person out'. As one man explained, 'they won't give him fire from their homes, they will not allow him to drink *chagga* with them'. Not being allowed to drink *chagga* with one's neighbours is considered a dreadful situation; not being able to get a light from a neighbour's fire when one's own has gone out is an inconvenience, but the most shameful consequence of being 'shut out' is that no one will bury the person when he or she dies (Hallpike, 1972: 65).

Occasionally the neighbourhood members work together on the fields for an individual. *Koda kanta* – 'neighbourhood work' – is called by someone who needs a large amount of work. I witnessed this only twice, when a woman requested two neighbourhoods to work for her on two separate days. Her household had fields in the lower part of Buso village, known as Karfura, where some of the fields are irrigable. She needed labour because the large *tomota* wall that usually protected this valuable irrigable land from flooding had been destroyed in heavy rains. By calling two neighbourhoods, Idigle and Nalaya, she demonstrated that it is possible to call people from neighbourhoods other than her own. In each case just over 100 people attended, and in order to encourage them to work she provided *chagga* for everyone, totalling four vats of *chagga* each time (a cost of around 200 *birr*[4]). She also gave each group 25 *birr*.

Koda kanta is rare, because few have the resources to make the *chagga* to provide the necessary hospitality and encouragement for the workers. But it is useful for those who can afford it, because it provides access to large amounts of labour. If a neighbourhood agrees to labour, all households of the neighbourhood are obliged to send someone to participate, and the same conditions apply to this work as to other neighbourly social responsibilities. The problems for a person requesting *koda kanta* are that sometimes the work is not done well, and that in return for calling people to labour on the land, those who have worked it are entitled to request and receive certain resources from the fields, for example, grass (for thatching), stones (for building), firewood, fodder, or the leaves of the *moringa* tree (for eating).

Clan

The clan or *kaffa* is the most important principle of social organization and primary signifier of identity. There are nine exogamous clans in Konso. Each person is born into a clan and takes the clan identity of his or her father. When a Konso person meets a stranger, one of the first questions is 'what is your clan?' Through this question the inquirer can quickly locate the stranger on his or her own mental map of social relations. Within one village, people of one clan are often related, but beyond this level it is not likely to be the case. Even so, shared clan identity can provide membership of an imagined community of relations above and beyond the village, and can provide access to other networks of trust and support.

The nine clans in Konso are Keertita, Paasanta, Tokmaleta, Eelayta, Mahalayta, Sawdatta, Isalayta, Arkamayta, Tikissayta (Shako, 1994: 81). Each clan is associated with certain wild animals, taboos and characteristics, but there is no hierarchy between the clans (Shako, 1994; Hallpike, 1972).[5] In addition to clan exogamy, in different areas of Konso there may be additional rules preventing marriage between some of these clans.[6] The clans are divided into particular lineages, also referred to as *kaffa*, and each lineage has a lineage head known as a *poqalla*, which is explored in more detail below.

Occasionally, rather like *koda kanta*, the clan can work together on someone's fields. *Koda kaffa* is called when a person is in severe difficulties. If a person is sick, or if there has been a death in the family, then the household might miss some crucial moments in the farming calendar and the year's crop could be ruined. This is not an uncommon scenario, as incapacitating illness and death are frequent. During the span of this fieldwork, there were epidemics of cholera and malaria. TB is endemic, and over the last decade, HIV/AIDS has become more common. Pneumonia, meningitis, kalazar (Visceral Leishmaniasis), and typhoid are all prevalent and result in hospitalization. For children, diarrhoeal diseases are the biggest killer.

If a family member dies then the family – including the household and relations by birth and marriage – are obliged to mourn formally. If an adult man has died, the immediate family is forbidden from working for nine days. If an adult woman has died, the family must not work for six days; if a child, for three days. During this time the family must stay in the house and receive relatives and friends who come to mourn and to offer their condolences. This ritual is called a *boyta* – a word derived from the Konso word for crying.

Thus *koda kaffa* – calling one's clan relatives to work together on the fields, or to help maintain a house – may be called at such a time of difficulty, or if a family needs assistance for other reasons. In this case, the brewing of *chagga* for the workers is the responsibility of the household. If they cannot supply *chagga* when they call for labour, there is an informal obligation for the family to invite the lineage to their house to drink *chagga* at some later date, when their troubles have passed. The number of people who come when *koda kaffa* is called depends on the size of the clan group in the village. In Buso, the largest lineage consists of over one hundred households, the smallest, only one or two.

Friends

Quite similar to *koda kaffa*, but much more common, is the calling of friends, relatives and neighbours to assist. Sometimes women may call women; at other times, men may call men, or the parties may be mixed. This is described as *fadeta*, and it implies asking for help and assistance. It may be derived from the root of the verb '*faD-*' which, according to Black and Shako, means to 'want, look for, find' and also to 'call for help' (Black and Shako, 1973: 30). This calling of friends to work together is common in Africa, and is described by Donham as a 'festive work group'. It is festive because the person who calls the people to work, must make sure there is enough *chagga* to drink, and preferably also beans to eat, to keep the workers happy. If a person has good connections, then perhaps 20–50 people will go to the fields for a household when they call. In such a case, to go for *fadeta* should be a pleasure, as people enjoy working together. But asking people to come and work for a person also implies a tie with the people who take part and an obligation to return the favour in the future, either by going to the fields or in some other way. No one is forced to go to the fields for *fadeta*, but not attending may also be interpreted as a statement that one does not wish to be sociable, and is not sympathetic, and it may have further consequences. There are few sources of help and assistance beyond those of friends, relatives and neighbours. As with *koda kaffa*, if a person is in extreme difficulties, he or she may call *fadeta* and yet not have the resources to provide *chagga*. In sympathy, people may assist, but without *chagga*, the workers will not work well, only staying for four or five hours in a day.

Parka and *marpara*

Parka and *marpara* are more formal kinds of collective work groups. They have fixed membership and they work cyclically on each others' fields, and also for non-members for money. They are usually made up of close friends or neighbours, and are often of people of about the same age, though in the case of younger groups an older man may be asked to join, to look after money collected and arrange the jobs. The main difference between *parka* and *marpara* is the size. *Parka* has generally around 10-20 members, whereas *marpara* has usually 20-50 members. *Parka* is derived from the Konso term meaning 'group', whereas *marpara* is derived from the Amharic term '*mahper*' meaning 'association'. *Marpara* is thus a quite recent innovation, and the aim of a *marpara* group is to raise money which is then lent out at high rates of interest (10 per cent per month). Each member of a *marpara* must also put a small amount of money into a kitty at the beginning of the season when the membership of the group is established for that year. Both *parka* and *marpara* sometimes charge their own members a fee, albeit reduced, to work on their fields. In general, a *parka* is less concerned with raising large amounts of money and more concerned with mutual help. They will work, at least once a year, on members' fields free of charge. *Parka* are often formed by young men and women, before they are married. *Marpara* groups are usually of men, though they have at least one female member, whose responsibility is to carry *chagga* and water to the fields, and to heat the water in the fields to add to the *chagga*. There are *parka* groups which consist only of women, and there are also groups which are mixed.

The payment for these groups varies greatly. Larger *marpara* groups are more expensive; those of young people may be cheaper. The payment can also depend on who is hiring, on his or her relationship with the group, and on the kind of work to be done. When work is requested it is described as either *koda dumaata* (a whole day's work) which means working from 8am till 6pm, or *koda murota* (meaning half) which would involve only four or five hours work. Whoever hires a *marpara* or *parka* must also make *chagga* for the workers, and, in order to get a good day's work from the group, should provide plenty of thick *chagga*.

These groups are organized to raise money and to work on each others' fields, but the money is rarely distributed back to the members. Instead, the group purchases an animal to slaughter in an annual feast at the end of the main cultivation season in September. This feast now takes place on the same day as the Christian Orthodox festival Meskerem, a day of feasting in celebration of the new year. The Konso festival has little to do with the Orthodox religion, but the fact that there is celebrating in all quarters at this time makes the Konso event feel more significant. It is a ceremony with a great deal of drinking, dancing and singing. The groups compete in terms of the beauty of the animals they buy for slaughter. The most successful groups slaughter an ox, the less successful ones a goat or sheep. The animal is stabbed with a knife through its spine at the back of the neck. Then it is stabbed through the heart. When this happens the animal screams loudly, and the people say it is laughing (*ihosali*). An animal that laughs long and loudly is thought to be a sign of good fortune. In his 1972 book, Hallpike describes this feast, and the blessings that accompanied the slaughtering. When I took part in this feast there were no blessings, but the slaughter was still carried out in a way to make the animal 'laugh', and the meat was distributed according to custom. Because the slaughtering is still considered to be a ritual act, the different groups have to be organized along religious lines. A person who has become a Protestant Christian will not eat the meat of a group where the elders have slaughtered an animal 'according to tradition'. Some Protestant Christians also have a problem with the singing and dancing that accompany the feasting. According to some of the Protestant Christian teachings, drinking alcohol is disapproved of, but as *chagga* is generally classified as a kind of food, the attitudes towards drinking it are relaxed.

In *parka* and *marpara* the practical, social and cultural dimensions of life that structure the organization of agricultural practice are blurred. At first glance, these work groups appear to be the most economically rational and oriented to getting the heavy work of terrace building and cultivation done. Further investigation reveals that they are engaged in raising funds to purchase an animal that signifies and celebrates their identity. A group basks in the reflected glory of their animal. It is as if the larger, more beautiful and stronger the ox they slaughter, the more powerful and beautiful they are themselves. In addition, the organization of the groups along religious lines shows that religious ideas, practices and beliefs (or discourses), which are often dismissed as insignificant and irrelevant to the practical task of sustaining a livelihood, are deeply implicated in organizing labour and networks of trust. The meanings, sometimes religious meanings, attached to different practices are central to motivating people to work together. They are a form of social capital. The cultural dimensions of agricultural practice are returned to in the next chapter.

Qoqopeta

Qoqopeta or *qopeta* is a new and still quite rare work group. Its name is derived from the Amharic term '*equb*' which describes the putting together of money in a group. These kinds of rotating credit or savings groups are common in other parts of Ethiopia. Here, members contribute their own money monthly, and they also work for payment. The money is also lent out to non-members of the group at a high rate of interest. The payment rates are the same as for *parka* and *marpara*, depending on the size of the group; the difference is that the members of the *qoqopeta* are generally interested in raising money, rather than buying an animal for slaughter. However, there is some overlap between these groups, particularly between the *qoqopeta* and the *marpara*. They also occasionally buy an animal to slaughter, and these feasts are looked forward to with great excitement. Following the slaughter of a sheep, the members of one group, made up entirely of older women, fix some of the white fat of the sheep's tail to their necklaces. They wear this even after it has become shrivelled and dry. It indicates how proud they are of their group and their productivity.

Koda pakaana

There is one other source of extra-household labour which is not strictly a work group. *Pakaana* is the name of a tuber[7] that grows wild in the fields. The plant flowers during the cultivation season, but its root is harvested during the dry season (January–February) which is also the hungry period. During this time, its location in the fields is difficult to find, as there is no surface growth. People who are short of food, therefore, offer to dig over the large fields of others, in order to find the *pakaana*. This is good for the field owner, as it turns over the soil, and the person who digs is allowed to keep all the *pakaana* he or she finds. The *pakaana*, which is caustic when fresh, must be dried in the sun and then ground before it can be made into dough balls (*dama*) for eating or making into beer. The preparation is awkward and time-consuming. It is a food used by those with little alternative.

Labour exchange?

In this intensive agricultural setting, the household is an important productive and consumptive unit. This is what might be expected following Netting's (1993) analysis of intensive production. But the existence of all these forms of organization, in which people work on each other's fields, demonstrates that the networks of cooperation and collective work produce the agriculture and the landscape. In the Konso case 'many hands make light work': when people go to the fields together work is completed more quickly than if a person was working alone. The labour needed for the intensive agriculture is produced by working together in ways that are socially meaningful: the neighbourhood is a residential network of support; the clan is made up of kinsfolk; *fadeta* calls on friends and relatives, and cements good relations through its activities. The cyclical work groups are also social groups with the shared goals of finishing the work and purchasing an animal for feasting. They often provide the opportunity for young men and women to spend time together. In 1996, the daughter of the family I lived with married a man from her *parka*. Such marriages resulting from friendships that begin in a *parka* are not uncommon.

As described in the introduction, these forms of cooperation are social, but they are also forms of institution which mobilize collective action, albeit in multiple ways. They are not as clearly defined, for example, as the institutions governing behaviour in irrigated agriculture, where they mobilize labour for 'operation and maintenance', distribute water, and mediate against conflict. But these institutions encourage co-operative work and may be just as important for the construction and continuity of the intensive agriculture.

The significance of the cooperative work parties is not just that they organize labour and get the work done. The implication is also that cooperative work is equal to *more* work. People may work harder together because they are giving each other encouragement, perhaps also new ideas, but is the work output really much increased by these methods? It is unlikely that this is the case with the extra effort generated by the social participation alone, but it is possible that it combines with other dynamics of household labour availability to make a significant difference.

In order to understand the dynamics of household labour availability it is useful to return to the work of Chayanov, as others have done (see also Sahlins, 1974; Donham, 1990). Chayanov demonstrated that, over time, households have different labour capacities depending on the number of workers in the household compared to the number of people that there are to be fed. Over time, 'peasant' households go through a development cycle: at some stages in their lifetime the number of mouths to feed (consumers) is higher than the number of workers; at other times, the number of workers compared to the number of consumers may have increased. For example, if a husband and wife have three children under the age of five, and an elderly relative living with them who can no longer work, then the dependency ratio (number of consumers/number of workers) is high. Later as the children grow and start to contribute to the work of the household, the dependency ratio decreases. When the dependency ratio is high, in order to produce enough to feed the household, those who can work must work harder.

For Chayanov, the power and enduring nature of the peasant household is its ability to withstand the pressure it is under when its dependency ratio is high. It has the potential to dig deeper, to work harder, when the workers are few and the mouths to feed are many. This ability is what Chayanov describes as the self-exploiting potential of the peasant household. But his work also demonstrates that it is possible for a household to be in a position where it has a relative labour surplus, when the dependency ratio is low. Some scholars, for example Sahlins (1974), have used this as evidence that the household producer is an inefficient form of production, as this surplus labour is under-used. But Donham's work in Maale showed that the exchange of labour that took place through 'working together', at least at one moment in time, meant that there was an exchange of labour between labour surplus and labour deficit households. Those with a low dependency ratio contributed more to 'working together' than those with a high dependency ratio (Donham, 1990). Over time, presumably, inequalities in the labour contributed to the community would even out.

In Konso, this is sometimes the case. Those with a low dependency ratio at times help out those with a high dependency ratio. It is evident that a high dependency ratio frequently emerges from situations other than the 'natural' development cycle of a household. It results from all too frequent sickness, or from the need to mourn a death in the family. Overall, contributing to the labour networks provides a way to cope at times when labour is short. In this way it is risk-averse, in a similar way to many of the agricultural techniques described in the previous chapter. Sometimes a household will contribute a lot of labour to working parties, while at other times it will receive it; a household works for others because it is impossible to know when the labour of others will be needed. A Konso saying goes that 'somebody who doesn't have anybody, tomorrow he will die',[8] demonstrating the need to invest labour in these networks of support.

Such an ideal interpretation of mutual help in Konso is attractive. The existence of an institution that incorporates a delayed reciprocity, in which the needs and abilities of different households are balanced out over time, would be invaluable for development. They might be a means to provide support for the most vulnerable. Such a mutual help institution conjures up the idea of a harmonious moral economy in which people redistribute their surplus capabilities and help each other out in times of need. It is a kind of 'primitive communism' in which labour is exchanged 'from each according to his abilities and to each according to his needs' (Sahlins, 1974: 94, quoting Marx). Thus this indigenous community is an equal one. The anthropological literature is rich with examples of levelling mechanisms which come into effect should someone accumulate more: Crewe and Harrison for example, describe how accusations of witchcraft followed a man's successful participation in a development project in Zambia. The fine he had to pay in order to compensate for his supposed witchcraft activities distributed the profits from the project to the rest of the community, and resolved any tensions that were growing from the new societal inequalities (Crewe and Harrison, 1998).

But though attractive, such equality must be proved; it cannot be assumed. Even a cursory review of the different kinds of work groups above reveals that although there may be an emphasis on participation and reciprocity, the use of the labour groups is characterized by differential abilities to call people to work on one's land. Any inequality in the exchanges does not necessarily balance out over time. On the contrary, the imbalances may be part of deeper, more structural inequalities in Konso society.

Factors restricting access to work group labour

It is clear from discussion of the work groups that the 'labour exchange' is not merely a matter of exchanging labour for labour, but that other goods are involved. The provision of *chagga* is requisite if good work is to be carried out, and the *chagga* that is brewed for others, like that for sale, is produced from grain purchased on the market. For many of the groups, some financial recompense for the labour is also required, even if it is only a token. Not everyone has the capacity, therefore, to call a work group.

A change over time in the nature of these institutional work groups may also be in evidence here. With the exception of *koda pakaana*, the work groups discussed above show increasing levels of formalization as one moves from the lineage, neighbourhood, and *fadeta* group work, towards the cyclical and paid work groups. The latter groups also show an increasingly commercial element, raising money, even if it is often used in the purchase of an animal for a ritual celebration. The increasing levels of commercialization may have some advantages: these groups loan money, sometimes reaching significant figures of hundreds of *birr*. One of the key arguments against indigenous agriculture is that the lack of formalized land titling means that money cannot be raised and credit cannot be obtained for investment (Bassett, 1993; Bruce, 1990; de Soto, 2000). This evidence demonstrates that credit is available through other means, without the mortgaging of land or other assets. But the levels of commercialization are also moving the work groups further from a mere exchange of labour for labour, and with them may also come an increasing inability of some people to raise the necessary financial capital to call the work groups.

A third form of evidence for the differential access to labour relates to the way in which some people have higher demands for labour than others, and this varies with factors other than the dependency ratio. The woman who called two large working parties (at least) was able to rebuild a flood protection wall, a large form of landesque capital. Not everyone has access to the fertile irrigable land that is protected by these walls, and not everyone has access to the labour to build the walls necessary to protect them from flooding. Some people are using more labour to cultivate their larger landholdings.

Differential access and use of labour

After carrying out more qualitative research and realizing that work parties were more than just occasional community activities to help out a friend or neighbour, I collected some more quantitative data about the use of work groups during the main cultivation season (March – September) in 1996. The findings of the quantitative research supported the existence of certain patterns that were observed before and since the survey. In a sample of 48 people, I found that the maximum number of person days of extra-household data was 476, and that the average number of person days received through work groups was 51. Four people in this sample, however, used significantly larger amounts of extra-household labour than the others, at 151, 230, 259 and 476 person days each. These four people alone had an average of 283 person days. The average among the rest was only 31 person days.

There was, therefore, a marked gap between the use of extra-household labour by these four people and the others in the village. This was not wholly unexpected, as this pattern had been described in qualitative interviews, and the sample was purposely selected in order to find out if it existed in practice. These four people were easily identifiable, not for their wealth, which these figures might suggest, but because they were all *poqalla*s, or lineage heads, people who have been described in previous literature as sacred leaders and priests (Hallpike, 1972; Amborn, 1989).

If the data of the two groups are disaggregated, it is possible to see that there is also a difference in the kind of extra-household labour that is received: the 44 people who were not *poqalla*s, received approximately 40 per cent of their extra-household labour through *fadeta*; approximately 41 per cent was received through paid work groups (such as *parka*, *marpara* or *qopeta*) in which someone in the household was a member; and the remaining 19 per cent of this labour was received from paid work groups in which no one from the household was a member (in other words, the labour was hired). For the four *poqalla*s, 47 per cent of extra-household labour was received through *fadeta*; 15 per cent through cyclical work groups in which one of the household was a member; 8 per cent was received from cyclical work groups in which no one in the household was a member; and 30 per cent was received from other kinds of work groups, including the neighbourhood and the lineage.

There was also a difference in the use of extra-household labour between the two groups. Those who were not *poqalla*s used the majority of labour they received to cope with peak labour demands at times of sowing and weeding. The *poqalla*s, in contrast, invested a higher proportion of their labour on activities such as terracing and the construction of other

Figure 2.1 *Percentage of extra-household labour allocated to each task*

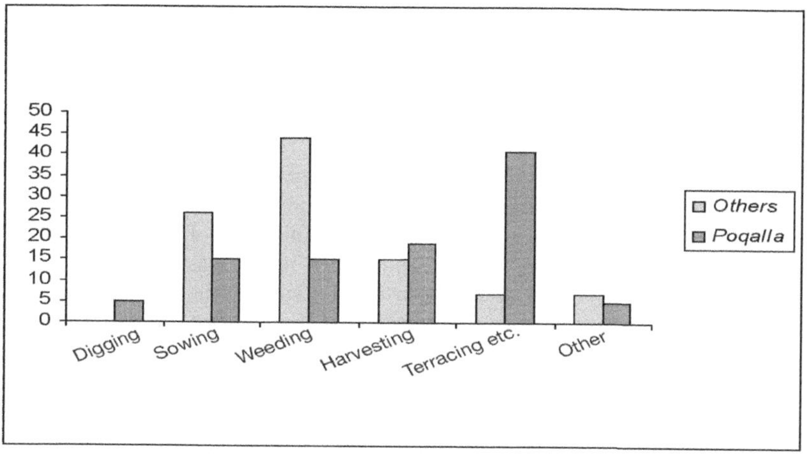

forms of landesque capital. They could, therefore be said to be investing labour in the construction of soil and water conservation structures and improving the land, whereas the others, are using their labour for more immediate subsistence needs (see Figure 2.1).

The evidence summarized here suggests that the *poqallas* are playing a role in the construction and maintenance of the intensive agriculture. This contrasts with previous views of *poqallas* as people holding a sacred position in the society, and also with previous ideas of the role and organization of group work in constructing the agriculture.

Amborn, for example, in his work on intensive agriculture of the Burji-Konso cluster, writes that the 'co-ordination of large agricultural and infrastructural measures, including the distribution of harvested goods on festive occasions' rests with a junior generation group, 'the "sons" – known as "warriors" in much of the literature' (Amborn, 1989: 80; 81). According to Amborn, after determining the layout of terraces and irrigation, this group recruits others to carry out the work. In his view, three different principles of social organization involved in the production of agriculture cross-cut each other and are in balance:

> The structure of the political system is thus characterised by factors designed to counteract tendencies of individual groups to gain dominance. The kinship system, the territorial units, the generation-grade group system and, with it, all offices of political and religious affairs interact to balance one another. Every individual belongs to each of these systems at the same time, and may hold offices in each, while being controlled by different groups and members representing distinct interests. In all three areas, there is enormous diversity allowing for a degree of flexibility that balances the

overall system even at times of extreme ecological, social and political stress (Amborn, 1989: 81).

In the past, Konso had a strong generation-grade system which organized all men into a particular generation grade in the way that Amborn describes.[9] As Amborn rightly states, however, the generation grade system in Konso has declined in recent years, a result of the impact of northern rule, the 1974 revolution, and the impact of religious conversion. It no longer functions in a management capacity, planning and executing the larger-scale infrastructural projects (if it ever did).

The data from Konso suggest that the organization of labour is a good deal more complex. They suggest that there never was one overseeing, organizing body that recruited labour from a single other form of organization. Work groups are not only organized by territory (for example the neighbourhood), but also by lineage and by friendship. These forms of labour are not in balance, and there is a good deal of inequality in access and use of labour. Theoretical insights from work on access to land have not viewed indigenous practices as harmonious and unchanging, but as the outcomes of dynamic power struggles of various kinds (Berry, 1989a; 1992; Bassett, 1993). Such struggles are integral to the form of the agriculture and are also catalysts for change within it. In Konso, the *poqallas*, positions that were previously understood as priests, have some role in structuring the uneven access to and use of labour. Through this, they play a key role in the organization and production of the agriculture and landscape. Their role in structuring access to labour can also be seen and understood further through an examination of the *poqallas*' access to land.

Land and labour

Studies of land and labour are frequently carried out independently of each other, but as Berry (1989a; 1992) and Bassett (1993) have indicated, the struggles over land and labour use and access are often closely connected. The *poqallas* have access to significantly larger amounts of land than others, and their land is among some of the most productive in the area. The *poqallas* need access to large amounts of extra-household labour in order to cultivate their larger landholdings, and they also use their control over land as a means of obtaining access to this extra-household labour.

It is not easy to know the extent of individual land-holdings in Konso, as people are hesitant to disclose what they possess (Hallpike, 1972). In the past this may have been because of a combination of fear of taxation, and also because counting up of landholdings is seen as an unattractive display of wealth and position (or a lack of it). This has been compounded by the fact that claims to land may be quite tenuous or contested, following the formal nationalization of land under the 1975 land reform programme.

The information in this study on land holding was collected by asking people where they had fields that they cultivated and by asking them to estimate the yield of grain obtained from each field 'in a good year'. This was measured in terms of *kolataas*, the skin bags that are used to bring grain home from the fields after harvest. This method was developed through talking with people about the way the value and productivity of a field is discussed emically. People seemed to reach agreement readily about these estimates and this suggested that it could be used as a means of comparison. As productivity varies depending on factors such as soil quality, water availability and condition of terraces and other forms of 'landesque capital', this seemed a more useful indicator of value than area. In order to be able to compare these figures with other data on land-holdings in terms of area, however, I measured a series of yields from particular fields and compared this with the estimated *kolataa* yields of those fields. By using a basic yield to area estimate, I was then able to calculate estimates of other land area sizes drawing on estimates of yields. Though imperfect, I believe this was better than using farmers' estimates of area size in hectares.

Despite the 1975 land reform that declared all land the property of the state and outlawed landholdings of more than 10 ha, customary claims to land have endured, and land is still considered to have 'its owner'. People distinguish between three kinds of land that they cultivate: land owned that was inherited from forefathers or relatives; land owned that had been bought; and land borrowed. These customary forms of landholding persist partly because there was no redistribution of land with the 1975 land reform programme. The above classification framework (inherited, bought, or borrowed) was used in this research to record information on the nature of the landholdings, with particular attention and openness to any alternative classifications or arrangements. In addition to this, those interviewed were asked if they had any fallow or forested land.

In Buso village, land-holding information was collected for 44 households who were not *poqallas* and for 25 *poqallas*. First the land that they cultivated was recorded. Of this combined sample, the mean land cultivated was 5 ha, and the maximum was 51 ha and the minimum was zero (n=3). This is significantly higher than expected. Existing land-holding information is rare, but figures collected by Messeret Lejebo (Messeret, 1990: 169) showed that for a sample of 372 people, 100 per cent had less than 4 ha of land, whereas in my sample, only 42 per cent had less than 4 ha, and 9 per cent had more than 10 ha (see Table 2.1 below).

The differences in these figures could be accounted for in various ways. Firstly, Messeret's figures are based on a larger sample than my own, and they might therefore be considered more representative. Secondly, they are also based on large-scale questionnaires which were carried out by enumerators, many of whom may have had contacts with, or be working in some capacity for, the local government administration. This is likely to have led to significant underestimation of land-holdings by informants who

Table 2.1 *Comparison of landholding data (Messeret figures are taken from Messeret, 1990: 169)*

	Messeret figures		My figures	
Area (ha)	No.	%	No.	%
0-0.25	77	21	3	4
0.26-0.5	115	31	2	3
0.5-0.99	91	24	9	13
1-1.25	39	10	6	9
1.26-1.5	22	6	3	4
1.51-1.99	13	3	5	7
2-2.99	8	2	8	12
3-3.99	7	2	4	6
4-4.99	-	-	9	13
5-5.99	-	-	5	7
6-6.99	-	-	1	1
7-7.99	-	-	5	7
8-8.99	-	-	1	1
9-9.99	-	-	1	1
10+	-	-	7	10
Totals	372	100	69	100

wished to protect themselves from taxation or re-distribution of land. Thirdly, Messeret's data is probably based on estimation of area size by informants, not the actual measurement of fields, and while I was in Konso I found that respondents had difficulty in estimating area. Finally, my sample data was biased in its composition, as it contained a high number of *poqallas*.

When the landholdings of the *poqallas* and those who are not *poqallas* are compared, it becomes clear that the amount of land cultivated by *poqalla* households is greater.[10] The mean land area cultivated by *poqallas* yielded 139 *kolataas*, and the mean land cultivated by others yielded 44 *kolataas*. For the *poqallas* the maximum yield was 815 *kolataas*, and the minimum was 13 *kolataas*; for the others the maximum was 217 *kolataas* and the minimum was 5 *kolataas* (excluding three interviewees who had no land). Putting it another way, 79 per cent of *poqallas* cultivated fields yielding more than 50 *kolataas*, and 73 per cent of the others cultivated land yielding less than 50 *kolataas*.

If the distribution of landholdings of *poqallas* is examined in more detail, it seems that the *poqalla* who cultivates land yielding 815 *kolataas* of grain is something of an exception. As well as his fields in Konso, he owns and cultivates a large amount of land in Gummayde, and he sells his grain in the nearby town of Arba Minch. His landholdings might also be seen to be so large that they make the mean of the *poqallas*' landholdings artificially high. However, even if he is removed from the sample, the

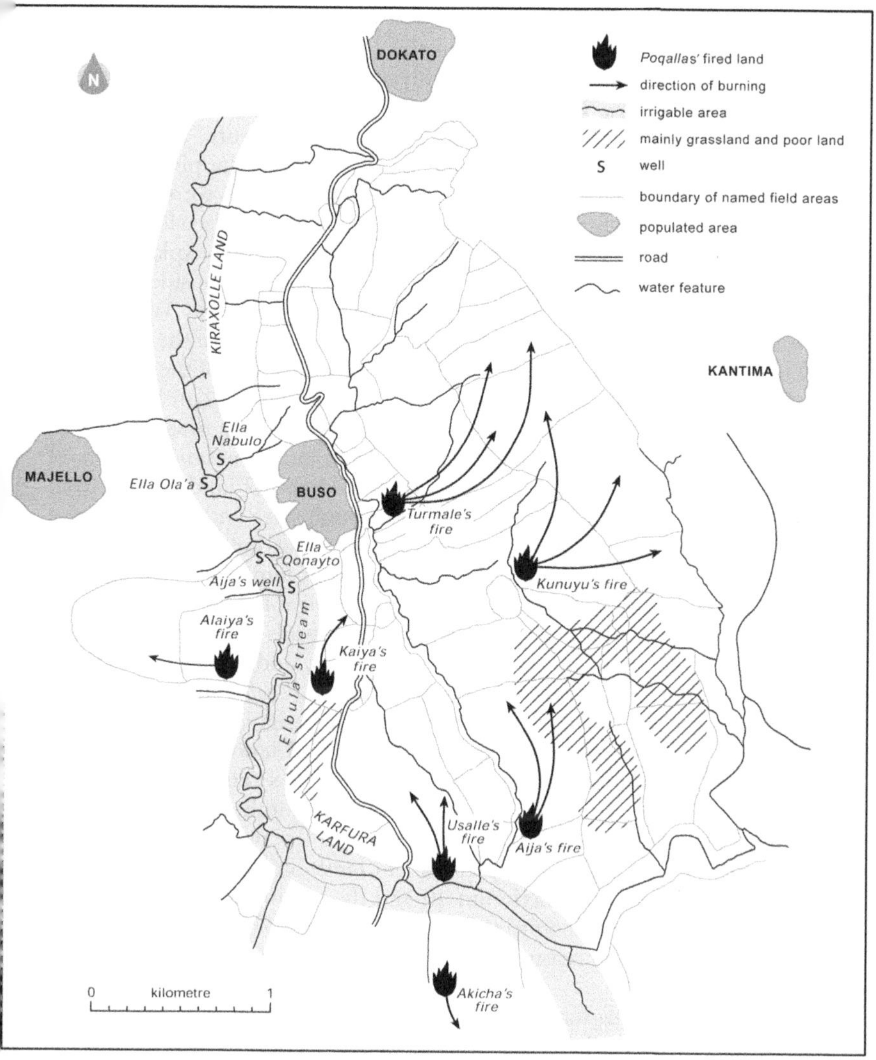

Map 2.1 *Buso field areas showing places where* poqallas' *fires were supposed to have been started and direction in which they burned*

mean yield is 110, still significantly higher than that of the non-*poqallas*.

The pattern of difference between the land cultivated by *poqallas* and others is clear, but these figures of cultivated land may only be part of a person's landholdings. They do not include fallow land, or land that has been lent to others. Fallow land is very difficult to measure, although it is often extensive, particularly in the case of *poqallas*. Mapping exercises found that areas of fallowed grassland surrounding villages were frequently thought to be owned by *poqalla* households, in Buso and elsewhere. Those people who took part in participatory mapping exercises explained how the *poqallas* had come to own this land: the *poqallas* are considered to be direct descendants of the first people to come to the area. When their ancestors arrived, they started a fire, and the place to where the fires were said to have burned marked out their land. An indication of the directions in which the fires are said to have burned in Buso village can be seen on Map 2.1. These are the areas where the *poqallas* still have grasslands, and also most of their cultivated landholdings. Over time the landholdings of these *poqallas* have been augmented with the addition of land bought from other people, or they have been depleted through sales and the division of land to descendants. In this way any homogeneous landholding 'blocks' have been broken up, but the general pattern is in some cases still evident. The map also shows the location of the good irrigable land, and the area known as Karfura, where large flood protection walls – *tomota* – are built.

Another difficulty in measuring the amount of land held by a person is that some people lend land to others to cultivate. *Poqallas* and others have land that they lend to people, usually to a friend, neighbour or relative. 40 per cent of non-*poqallas* who were interviewed had land that they lent to others, but the amount of land they lent was extremely limited: the majority lent only one field, and the maximum number of fields a non-*poqalla* had lent was three. This is in contrast to the number of fields lent by *poqallas*: only 48 per cent of *poqallas* lent land, but the average number of fields lent was 15, the maximum 46 and the minimum 1. If the area of land lent is taken into account, then the mean landholding yield of *poqallas* can be estimated as increasing from 139 to 180 *kolataas* (an increase of 29 per cent).

The pattern in this data is strong. In land cultivated, in land fallowed, and in land lent to others, the *poqallas* are significantly 'richer' than the others. Not all the *poqallas* were very 'rich', and this is explored in more detail later, but, overall, there was a connection between *poqalla* status and wealth in land.

These large landholdings are related to the *poqalla*'s greater access to extra-household labour, as land lent to others is given in return for labour. Customarily, the land lent to others by a *poqalla* is called *piyolada*. In return for *piyolada* land, *piyolada* labour is given. The person borrowing the land must go to work on the fields of the *poqalla* three times a year: once at sowing; once at weeding; and once at harvest. Since the 1975 land reform, *piyolada* has been outlawed. The revolutionary government viewed it as an extractive and exploitative practice by large landowners who should, in any case, no longer be in control of such large amounts of land. Anyone requesting this *corvée* labour should be punished by a fine or imprisonment, assuming that they had not already been imprisoned for what was considered to be their 'exploitation' of the people in pre-revolutionary times. The history of this process and the impact that it has had on the position of the *poqalla* is explored in more detail in Chapter Six. Here it is necessary to confirm only that, in the mid-1990s, the data collected demonstrated that some *poqallas* still controlled large amounts of land, and that some people were labouring on the *poqallas*' fields in return for the land that they received as a loan. In order to avoid legal sanction, however, the land and the labour were no longer called *piyolada;* instead, it was simply called *fadeta,* and the work carried out was no longer as formalized as three times a year.

The case of one *poqalla* can be used to illustrate *piyolada* in more detail. He lent land to people in his own lineage, and also to others. His lineage was made up of 23 households, and he lent land to 7 of these. He lent land to 17 other people, some of whom were relatives by marriage; others were neighbours; others were simply described as friends. Over three years, 1994–96, the *poqalla* called a large amount of extra-household labour which was described as *fadeta.* Those who contributed to this labour were made up largely of people who were of his lineage or who had borrowed land. The information is summarized in Table 2.2 below.

This table shows that the lineage members performed work for the *poqalla,* but that those who borrowed land provided slightly more labour, particularly in years when a high amount of labour was required. Those who were not members of the lineage but who borrowed land also

Table 2.2 *Amount of labour carried out by different groups on the* poqalla*'s fields*

Group	Labour (person days)	1994	1995	1996
All lineage members (n =23)	Total	137	24	41
	Percentage of all labour	62%	32%	47%
	Average per person	6.0	1.0	1.8
Lineage members who do not borrow land (n = 7)	Total	39	10	11
	Percentage of all labour	18%	14%	13%
	Average per person	5.6	1.4	1.6
Lineage members who do borrow land (n = 16)	Total	98	14	30
	Percentage of all labour	44%	19%	34%
	Average per person	6.1	0.9	1.9
People who borrow land who are not members of the lineage (n = 17)	Total	84	50	46
	Percentage of all labour	38%	68%	53%
	Average per person	5.6	2.9	2.7
Total (n=40)	Total	221	74	87
	Average per person	5.5	1.8	2.2

contributed significant amounts of labour. It demonstrates the variation in the amount of labour that is needed from year to year. This *poqalla* also has lands in the fertile irrigable valley area below Buso village, and in 1994 his flood protection was destroyed and needed to be rebuilt. Hence he needed a large amount of labour. 1995 was a dry year, and the need for labour was low. The data show that although *piyolada* labour no longer exists in name, and labour is no longer given in an institutionalized form three times a year, it still exists in the sense that when the person who lends land to them calls a *fadeta,* they are obliged to attend. This labour is particularly valuable when a large amount of labour is needed. As it now takes the form of *fadeta,* and is discussed in terms of assistance, it is difficult for an outside body to prevent people working in this way, even if it wished to do so.

The much-needed labour for intensive agriculture is generated therefore by the many social networks that exist between households. These forms of social capital are vital for the production and reproduction of the agriculture and the landscape. But the power relations that exist within these forms of social capital mean that they cannot be considered as neutral and cannot be harnessed for development without repercussions. It is not necessarily possible to use them instrumentally to help everyone, especially those who are seen here to be the poorest and most powerless. The participation in the networks of cooperation and land and labour management is not equal, nor is the receipt of the benefits from these institutions. The organization of the labour and the way in which it is divided between people is influenced by the position of the *poqalla.* The data presented in this chapter illustrate that *poqallas* are forms of institution who have power and position and a large share of the resources. They have a larger share of land than others, and they use this share of land to gain access to a larger share of labour. The labour they have access to is used in the construction of the landesque capital, that improves the land capability and enables a larger population to be supported. In the next two chapters, I look in more detail at the position of the *poqalla,* and its significance for development and for understandings of indigenous institutions.

Notes

1 To make *chagga,* the flour is mixed with water and fermented in large wooden vats, carved from single pieces of wood. Only some women possess these vats and they may lend them to others in return for help with the brewing or serving of the beer. Some networks of cooperation exist between women to share vats and work, and to make sure that they do not all sell beer on the same day; in such a case the amount of beer might outstrip demand, and might not be sold. It does not keep for more than a day or two.

2 Nalaya, Idigle, Tarajo, Kolalta and Kuile. Kuile is now small and is combined with

Kolalta for some administrative purposes. Another area, Iffa, sometimes acts as a neighbourhood, but the residents are more commonly incorporated with those of Nalaya, Idigle or Tarajo.

3 In contrast, women frequently cross *kanta* borders when they marry.

4 In the mid-1990s, 1 GBP was equivalent to approximately 8 *birr*. At the beginning of 2008, 1GBP was equivalent to 13 *birr*.

5 Sawdata, for example, the clan of the family I lived with in Buso, and which was extended to include me, is associated, perhaps unglamorously, with the baboon; Argammayta is associated with a camel; Mahalayta with a hyena; and Passanta with a fox.

6 For example, in Karate, Keertita and Paasanta cannot intermarry; in some areas of Fasha, Arkamayta and Isalayta cannot intermarry. These correspond to so-called clan alliances that were described by Hallpike (1972: 88) which were accepted by Shako (1994: 81) (Keertita-Paasanta, Tokmalayta-Eelayta, Mahalayta-Sawdata, Arkamayta-Isalayta, Tikissayta), but there is no full division of Konso into different moieties as there is in Boran and many other southern Ethiopian groups. There are further location-specific arrangements, however, where different clans cannot intermarry, which do not follow the above pattern. For example, in the area of Iffa in Buso, I found that certain lineages of Sawdata cannot marry with certain lineages of Arkamayta. Shako notes also how in certain areas, some sub-clans of the Kerrtita clan can marry with each other (Shako, 1994, ibid.). The clan unit is therefore not as inflexible and determinate a unit as might be first assumed, but can be contextually and locationally varying. These are, however, modifications to a general pattern.

7 *Stylochiton hypogeum, Sauromatum nubicum*, or *Amorphophallus abyssinus*, Yohannes *et al.*, 1996.

8 The saying is *'nama a nama qapo namakulayta'e to'ni'*, which translates literally as 'someone who has someone else dies the day after the day of tomorrow'. The saying implies that a person who has support can survive longer than somebody who has no support. A person without support may die very quickly. I am grateful to Ongaye Oda for supplying an accurate translation.

9 In Karate region of Konso all lineages were divided into two generation categories: *hiriba* and *kalgussa*. The generation grades passed through a series of stages, alternating from *hiriba* to *kalgussa*. Fathers and sons would all be either *hiriba* or *kalgussa*, and thus they were always two generation stages apart (Hallpike, 1972).

10 Here some of the possible errors in this way of calculating land area are reduced as the same method of estimating land size and value is used with both groups.

Three

Ritual Life of Agriculture

> The world becomes apprehensible as a world, as cosmos, in the measure in which it reveals itself as a sacred world (Eliade, 1957: 64).

The *poqallas*' control over larger amounts of land gives them access to more labour, which allows them to invest in the construction of the landesque capital. As institutions, the *poqallas* are structures of power, and the material presented in the previous chapter suggests that their power is derived mainly from their economic control, and that a *poqalla* can be perceived as a kind of landlord. The revolutionary *Derg* government certainly took this view when it assumed power in 1974, and accused many *poqallas* of being landlords who exploited the people. The extent to which the *poqallas* can be justifiably viewed as landlords is examined in more detail in Chapter Four; Chapter Six looks at the way in which these government assumptions impacted upon the position and role of the *poqalla* after the revolution. In this chapter, I examine the nature of the different exchanges between the *poqallas* and others in the community and the way in which they go far beyond the realm of the economic. In the literature, the *poqallas* have first and foremost been understood as sacred ritual leaders or priests, and it is this aspect I consider here. This chapter argues that the ritual role of the *poqalla* is as important to the production of Konso landscapes and livelihoods as the economic role, and that the ritual and the economic cannot be separated. This has implications for understanding the nature of power relations in Konso, and for understanding the agriculture, how it is produced, and the potential that it has for the future. It also has wider implications for understanding the potential that indigenous institutions have for development.

Those who have written about Konso have varied slightly in their interpretation and translation of the position of *poqalla*, but most have stressed its religious and political dimensions. Amborn, for example, describes them as 'sacred chiefs' whose influence 'in the first decades of this century remained totally limited to the sphere of religion. They were, however, highly respected as mediators between hostile towns' (Amborn, 1988: 754). Shako describes *poqallas* as clan leaders. By doing so he emphasizes the way in which they have responsibilities to their clan members, as well as to the wider community. Their roles are summarized as 'blessing, advice, dispute settlement and reconciliation of warring towns' (Shako 1994: 85). Shako emphasizes their role in dispute reconciliation between clan members, as well as between villages. For Hallpike, the *poqallas* are 'priests' whose 'essential function' is 'to bring Life and peace (*sic*)' (Hallpike, 1972: 43). The appropriateness of the label 'priests' is questionable, given the way in which the *poqallas* have many different roles, but the short phrase 'Life and peace' does embody, effectively and efficiently, the most important dimensions of their position. 'Life' refers to the way in which, in the minds of Konso people, the *poqallas* represent their origins. It also refers to a person's health, well-being and prosperity: 'Life' is fertility in the broadest sense, in terms of fecundity, producing abundance, and being fruitful. It relates to crops and livestock as much as it does to people, as their fortunes are interconnected. This ability to 'bring Life' to others is the sacred and religious dimension of the *poqalla*. The second dimension, to bring peace, is seen in the *poqallas*' role in resolving disputes between individuals and between towns. Peace and prosperity are connected in Konso minds: without peace, misfortunes will affect everyone and everything; without peace, sickness will come and rain will not fall. In short, the fertility described above will be spoilt. These political and religious dimensions of the role of the *poqalla* are interconnected. In addition, as will be shown, they are connected to the hitherto unrecognized role that the *poqallas* have in the management of land and labour. This demonstrates that politics, meaning and economics are interconnected and shape the nature of agriculture and livelihoods (Shipton and Goheen, 1992). *Poqallas* are not the only important 'functionaries' in Konso society, but they are central, and an exploration of their position provides a prism through which broader issues and social relations can be examined.

The study of the ritual roles of the *poqallas* also has much wider resonance. They resemble the characters in myth that embody the fertility and wellbeing of people and nature. These are the sacred leaders and divine kings of Frazer's *Golden Bough* and Jessie L. Weston's *From Ritual to Romance*. Through the myth of the Fisher King, Eliot's *The Waste Land* explores a society in which the harmony of nature and culture is lost to modern life and cynicism. Scholarly interest in these institutions has languished as investigators have been afraid of viewing non-Western cultural forms in terms of a Western mythological prototype, distorting and misunderstanding them. Frazer's own work was dismissed for this kind of universalizing, combined with over-active editing and sensationalization of his subjects. However, one consequence of a more cautious approach to religion and ritual has been a lack of attention to the relationship between ritual leaders and their environment. In addition, any commonalities that exist between leaders such as the *poqalla*s and others leaders elsewhere have been ignored. One exception to this trend is Schoffeleer's (1979) collection of papers which examined territorial fertility cults in Central Africa, and the way beliefs in certain ideas and leaders are connected to certain kinds of environmental management. More recently, there have been few studies of the religious and the natural and the way they inter-relate. For Amanor, '[w]hile such rituals [which designate certain aspects of the environment sacred] may protect the environment, no environmental consciousness is articulated with them... [An approach to understanding] through "custom" will therefore not get us very far' (2002: 127–8). Here, I aim to demonstrate that religious beliefs, religious institutions and ritual practices, and the meanings attached to them, are highly significant in structuring the mundane matters of agricultural practice and landscape production. Even if they do not overtly articulate a message of environmental conservation, they may still be relevant. The religious dimensions of life can and should be studied.

Religion is far from irrelevant or superfluous to the 'real business' of agriculture and economics. Where religion is a part of everyday life, it structures experience and personhood. It governs what is legitimate and illegitimate in human behaviour and practice. As such it can be viewed as a powerful discourse and institution. Religion legitimizes hierarchies and forms of authority and is central to power relations. For some scholars, its power comes from its ethos that provides the regulations, codes of behaviour and ethics by which people live. For most anthropologists, it is people's worldview, the cosmology and logic by which people live (following Durkheim 1965, and Geertz 1973). It permeates every aspect of life, including one's very sense of self and place in the world (Delaney 1991). Turner's work stresses the emotional and experiential content of religious belief and practice, which adds an immediacy and power to religion and ritual (Turner, 1969). All the approaches emphasize that the influence of religious belief and practice stretches far beyond the spiritual. Here, we see the influence of Eliade (1957), who contended that thinking in terms of a spiritual/profane dichotomy is misleading. For Eliade (1957) the pervasive influence of the spiritual was as evident in life in the 'modern' 'West' as it is in communities like that of Konso. For Ingold (2000) and Delaney (1991), separating the religious from the everyday matters of agriculture and livelihoods is to impose a nature–culture dichotomy that may not be appropriate. All these approaches point to the need to take the religious and ritual dimensions of life more seriously, and to explore their influence on the practical aspects of life, typically more the concern of development thinkers. First, however, we must look at who these ritual leaders in Konso are, and examine the nature of their power.

Poqallas: Myths of origin and power

Poqallas are all male. Their position is inherited and derived from their claims to be the direct descendants of the first people to come to Konso, and they are thus considered to have given birth to everyone else. As Shako explains, '*poqalla* families were among the first ancient immigrants to Konso and are believed to have begotten the rest of Konso' (Shako, 1994: 82). 'Being first' denotes seniority, but it is a relative condition. It is possible to be born first in Konso, or in a village, or of a clan, or of a generation, as well as in other situations. The most important and most powerful kind of *poqalla* is thought, generally, to have arrived before anyone else. His position and power is evident in the way in which he lives separately from other people. This kind of *poqalla* lives in large but isolated homesteads, away from the densely packed villages, often on the top of hillsides. Spatial separation in this case indicates and reinforces their power. It is said that these *poqallas* cannot live among others because it is feared that their power would 'burn the people'.

Table 3.1 *Regional* poqallas *(Shako, 1994, 82)*[1,2]

Poqalla family	Clan	Residence
Kala	Kantayta	Kamoole
Kalmale	Tokmaleta	Itikle
Quufa	Keertitta	Kaaho
Daxasha	Tokomaleta	Pattankaldo
Pammale	Arkamayta	Dekatto
Ikta	Iktayta	Paayti
Afarta	Ishalayta	Timpalla
Orkeeta	Tikissayta	Kasarkiyo
Akkole	Sellata	Kuyle
Loxopa	Paasanta	Ala
Sirto	Keertitta	Tarra
Xolo	Mahaleta	Ala
Titiipa[3]	Sawdatta	Gidole
Olle	Keertitta	Kawwadi
Alkama	Keertitta	Xolmi
Marrota	Keertitta	Teepana
Akoole	Tokmaleta	Toxa
Olmoke	Tokmaleta	Faashi
Xola	Tikissayta	Extinct

These *poqallas* are described by Shako as 'clan leaders' and by Hallpike as 'regional priests'. Their position is marked by wearing a thick whitish metal bracelet known as a *tuma*. Their position in some ways is territorial, as each of these *poqallas* has villages that show respect and allegiance to him. According to Shako, there are eighteen regional *poqallas*. He lists them in a table together with their clans and the village nearest to their homesteads (see Table 3.1).

In Konso, explanations for the importance, power and responsibilities of regional *poqallas* are usually given in terms of the origin myths of these *poqallas*. These myths are narrative devices with explanatory power that legitimize certain power relations and practices (Fortmann, 1995). The origin myths relating to *poqallas* are many and varied, and relate to different aspects of their position and their relations to others.

Many of the origin stories discuss the *poqalla*'s status as first-comer, and several *poqallas* are considered autochthonous (Hallpike, 1972), to have literally sprung from the earth. Bamalle, for example, one of the two most senior *poqallas* in Karate region, is thought to have been born from a branch of a large sycamore fig tree when it was blown from a tree in a high wind. Five small male children emerged from the branch: Bamalle, Orkeeta, Afarta, and two others who are now unknown, but who reputedly became similar leaders in Boran. Bamalle was the first to emerge and the first to cry out, and so he was considered senior. The others became regional *poqallas* at Kaserkiyo (or Gazergiyo) and Timpalla (in Turo).

Similarly, Marrota's ancestor is thought to have been born from a gourd. The gourd grew to a large size and finally broke, revealing a baby inside it. In the baby's hand was a sprig of teff, showing that he was a farmer (*etenta*).[4] In due time the baby grew and gave birth to many descendants. The story of birth in a gourd is a more common one than that of birth in a tree. These are not merely remote ideas from distant times: I was told that a gourd was found in 1986 that was growing to a tremendous size, and people realized that it contained two new *poqallas* who were about to be born. It was said that, as the people also felt that there were enough *poqallas*, they took the gourd and threw it into the Sagan River.

According to these stories, the *poqallas*' position appears to be a classic example of deriving seniority through 'firstness'. This common principle has been explored in some depth by Kopytoff. For Kopytoff, the structuring principle of 'firstness' is derived from the frontier dynamics of much of Africa in which priority rights are established by first occupation or first use. This situation develops into a general principle of firstness that pervades much of life, power relations and values:

> The general value of 'being first' appeared in many different contexts in Africa. One could be the first-born among siblings, the first wife in a compound, the first in age in the lineage, the first entrant into a ritual

> initiation, the first in a region to acquire a title, the first village to have set up a secret society or a cult house. All these were ways of being senior, of enjoying precedence in the etiquette of public events and, not least, of being remembered after one's death (Kopytoff, 1987: 22).

However, not all regional *poqalla*s are considered to be either autochthonous or firstcomers. The origin stories of other *poqalla*s stress the way in which they are special, and, if not set apart from other people through birth itself, then through their special powers. These stories follow another archetypal pattern: that described as the 'stranger king', a man who comes from outside the community, who has special powers, helps the community, and who becomes its leader.[5] Guufa, the most senior *poqalla* in Fasha region, is a good example. He supposedly came to Konso at a time when the earth in Konso had opened up as if from a great earthquake. The earth split causing many deep gullies. The people who were seeking a solution to this problem invited Guufa, who was then a *poqalla* in Ilota (Gummayde, to the north of Konso) to come and help. One account of the history continues:

> Guufa came to Konso, and he asked the people if there had been other *poqallas* there who had tried before him and failed. He said he would try but he made everyone agree that if he succeeded then he would be the most senior *poqalla* of all these *poqallas*. The people agreed. Then he went to the area of gullied land and put the feet of an elephant over his own feet. Then he walked over the land. As he walked the land shook and the cracks came together and said 'Nyek' as they sealed. But there is an area of terrible gullies and erosion near Gazergiyo that he left as a reminder to the people of what he had done. Then the people respected him and he became the most senior *poqalla* in the area [Fasha].[6]

In this story, Guufa heals the ground that has become eroded and makes it cultivable again. The area of erosion, which Guufa is said to have left as a reminder, is today referred to as 'New York', as the eroded land takes the form of high buildings. It is a popular destination to go to take photos (particularly on marriage) and for tourists.

The regional *poqalla* Kalla, the other most senior *poqalla* in Karate region, also fits this pattern of 'stranger king'. He supposedly came from the north, from a place called Kontari. When he came, it is said that he could already predict the future and give advice. When he slept, he was visited in his dreams by a spirit who told him what would happen. According to his story, when he came to Konso, people saw that 'when he prayed to God, God listened to him'. There was no rain at that time because, it was said, 'God has forgotten the people'. Kalla prayed to God, and it rained.

A final element of the origin stories stresses not so much the *poqalla*s' first-comer status, but a related idea of bringing culture and order to the people. Some for example, describe how when the regional *poqalla*s came to the area, the people were 'just living in the forest', 'they didn't know anything', and 'they didn't have clans or villages'. Certain accounts of Kalla's history follow this pattern and describe a big meeting that was held at the beginning of Konso history, in which Kalla allocated people to villages and clans. When newcomers subsequently also came to Konso, they supposedly went to Kalla's home, where he again allocated them to a village and a clan. In one story of the history of Bamalle, Bamalle brought the drum to the people. The drum is a symbol of law in Konso, and hence he also brought order:

> Long ago, when the Bamalle first came to Konso, there were people coming from all over, from Sidamo, Gofa, Goria [Chench'a] and so many other places. It was just forest then and people lived in the forest. No one knew who else lived there, they could just see their fires. Lions lived in the forest. People were very poor and had nothing to eat or drink. Then the people came together with Bamalle and made the drum, and they played the drum and every one came to that place and prayed to *Waqa* (God) and called for rain. After that it rained and people in the forest began to till the land and raise crops.[7]

In these accounts, the *poqalla*s are bringing order to previously chaotic lives. In these stories, the *poqalla*s bring good rains and good lives by establishing certain cultural principles. The cultural principles or traditions are described in Konso as *aatta*, and it is said that the *poqalla*s keep the *aatta*. It is through keeping the *aatta* that 'Life' is maintained. Kopytoff also describes this form of myth as a rhetorical claim to firstness: 'By claiming that the land was wild, uncivilized and disorderly, they can present themselves as the firstcomers that have brought order to it' (Kopytoff, 1987: 121).[8]

Many of the different oral histories told are contradictory, but all are narrative devices for explaining the seniority of the regional *poqalla*s, their powers, and their relations to others in the community. Some of these stories fit with an idea of people coming from all over the area, and 'becoming Konso' – something that has been suggested in the study of the labour-intensive agriculture. These stories are powerful. They set the *poqalla*s apart as special in various ways, but they also establish connections between them and the rest of the people, connections which are of real or fictional kin.[9] The stories also establish the *poqalla*'s role in managing nature. The stories should not be seen as the historical truth but as part of a process in which different versions legitimize different claims and positions. All *poqalla*s are first, but some are more first than others.

The regional *poqalla*s are the most important form of *poqalla*, and they are socially and spatially relatively distant from most Konso people. In contrast, there are second order *poqalla*s, whom I refer to here as local *poqalla*s, and who are more closely connected to the people, and who live among them in the villages. Shako describes these *poqalla*s as lineage heads. Hallpike describes them as *mora* or lineage priests. *Mora* refers to the sacred and public meeting places that exist in the towns and which some of these priests preside over. Lineage priests refers to the way in which these *poqallas* are the heads of more localized lineage branches.

The origin stories of these local *poqallas* connect them to the regional *poqallas* in that the local *poqallas* are thought to be the descendants of the regional *poqallas*. In local histories of early Konso, the regional *poqallas* are portrayed as more mobile, and in each place in which they stayed for a period, they had sons. They told each son to be a *poqalla* and to look after the people in his birthplace. An account of the early history of Kalla, recorded by three elders, explains:

> Kalla came from Kontari and he went to Burji in Sidamo. There he gave birth to Uumi who is *poqalla* there. He left Burji and went to a place called Girosi in D'irashe. There he gave birth to Dahe, who became *poqalla* there. Then he went to Turo, where he gave birth to Tumtakallalayle, who became *poqalla* there. Then he went to Fasha, to the village of Kashale, where he gave birth to Yantale, who became *poqalla* there. Then he went to Gera, where he gave birth to Kwila, who became *poqalla* there. Then he went to Buso, where he gave birth to Ola'a, who became *poqalla* there. Then he went to Nalaya Sagan and Kantima, where he gave birth to Kaybo, who became *poqalla* there. Then he went to Kuili, where he gave birth to Akole, who became *poqalla* of Jarso. Then he went to Dokato, where he bore Tiita, who became *poqalla* of Hulme. Then he went to Gamole, where he gave birth to Maduro, who became *poqalla* there. Then they went to Farra mountain near Gamole, the place where they still live. They wanted also to leave this place, but all the people united and said that Kalla must not leave Konso. They planted a grass called *dimaata*, which grows flat along the ground, and from that day, Kalla settled.[10]

Although a written account (it had been dictated and was hand-written on loose paper), this demonstrates well the rhythmic and repetitive nature of oral histories. It is similar to other accounts of oral histories of movement and historical events in the region (see for example, Schlee, 1992) and to Biblical accounts of the movement and origins of people.

Through the telling of these stories, these second order *poqallas* are established as firstborn in their own right. All the rest of the people in the village are believed to have descended from them. A connection is thus made between the regional *poqallas* and the local *poqallas*, and with the descendants of the local *poqallas*. The relations between them become understood as ties of blood.

These local *poqallas* are the same *poqallas* discussed in the previous chapter, who owned large amounts of land and who used their land to obtain access to more labour. Their claims to land came from their claims to be the first-comers, a common aspect of the first-comer principle (Kopytoff, 1987; Murphy and Bledsoe, 1987; Fortmann, 1995). As we have heard, when these *poqallas* settled in the area, they lit a fire and the area over which the fire extended became their land. Regional *poqallas* also have large areas of land, or have had in the past. They also receive labour from others, which is particularly necessary because regional *poqallas* are forbidden by their sacred position from cultivating fields themselves. The regional *poqallas*' fields are more dispersed than those of the local *poqallas* and those involved in borrowing land and labouring are from more disparate places. It was, therefore, more difficult for me to collect systematic quantitative data on these practices than it was for the local *poqallas*, whose activities are limited generally to one village. The observations of this research tended to show, however, that the arrangements of the regional *poqallas* over land and labour mirrored at a larger scale those of the local *poqallas*.

At a village level, there is a process by which a new *poqalla* can be formed. This is connected to the division of a lineage into two branches, each with its own lineage head (local *poqalla*). The Konso explain this process in relation to bees in a hive. They say when the bees in a hive

Table 3.2 *Local* poqallas

Poqalla	Clan	Lineage size (number of households)	Major or minor
Aija	Sawdata	106	Major
Akicha	Argammayta	12	Major
Alaiya	Tigisaita	9	Major
Aysheki	Argammayta	8	Major (descended from Bamalle)
Ba'ayto	Sawdata	38	Major
Balde	Sawdata	29	Major
Doxole	Keertita	50	Major (descended from Kalla)
Jemolla	Argammayta	10	Minor (descended from Kucho)
Kaiya	Tigisaita	76	Major
Kuch'o	Argammayta	51	Major
Kulili	Argammayta	13	Minor (descended from Kucho)
Maxo	Argammayta	0	Major
Nabulo	Keertita	2	Major (descended from Kalla)
Ola'a	Keertita	56	Major (descended from Kalla)
Onaka	Passanta	19	Major
Otcha	Argammayta	23	Minor (descended from Turmale)
Qonayto	Keertita	14	Major (descended from Qonayto)
Salomo	Argammayta	8	Major (descended from Kalla)
Sima	Tigisaita	33	Major
Tarafallo	Keertita	5	Major (descended from Kalla)
Towaysha	Argammayta	32	Major
Turmale	Argammayta	52	Major
Usalle	Argammayta	30	Major
Worrarra	Sawdata	24	Minor (descended from Ba'ayto)
Xaiyko	Tigisaita	39	Major
Xelle	Argammayta	22	Major (descended from Bamalle)
Xombollo	Sawdata	25	Major

become too many, it is necessary for some of them to break off from the first hive and go somewhere else under a new leader. In the same way, if a lineage becomes too many, then a new *poqalla* should be blessed, and be the leader for his descendants, for his branch of the lineage. Strictly, the new *poqalla* should be the descendant of the eldest son of a second wife, who is called 'Lamita'. This process has been described by Hallpike: after five generations, this son can become a *poqalla*, if he is blessed and given a knife by his own *poqalla* to use to slaughter animals in ritual (1972: 101).[11] The five generations are symbolized by the five iron bracelets known as *mugla*, worn by the inaugurated local *poqalla*. However, in practice the acceptable number of generations is variable; it depends on the individual's ability to amass capital in order to fulfil the celebrations and rituals that are necessary for his inauguration, and on whether or not he is blessed by the original *poqalla*. After a new *poqalla* is blessed, in various situations he is considered less important than the original *poqalla*, and this seniority is made clear in certain rituals.[12] Where it is necessary to make a distinction, these less important local *poqalla*s will be termed minor local *poqalla*s because they represent a minor lineage segment. Similarly, the first more important *poqalla*s will be referred to as major local *poqalla*s.

In Buso, a village of approximately 750 households, there were 28 established local *poqalla*s. Table 3.2 shows their names, their clan and their residence area in Buso. It also shows the size of their lineage, which means the number of households who consider themselves to be descended from that *poqalla*.

One of these families (Maxo) was extinct, as the family and their lineage had mostly died out. Tarafallo and Xombollo were also in difficulties, because, although there were still lineage members, the *poqalla* line had died out. This withering of a *poqalla* line and lineage is considered tragic, as the *poqalla* has not been able to fulfil what Hallpike describes as 'his essential duty ... to bring Life [*sic*] – health, fertility, and peace – to its [the lineage's] members' (1972: 99). Without a *poqalla* preventing quarrels and conflict and performing certain ceremonies, 'a lineage... will suffer disease, its crops will fail, its women and cattle will become sterile and it will die out' (ibid.). Two men describing such a family said 'the house is reduced (*helumi*), they are weak like children, they can't make food, they don't have any grain in the granary. Terrible!'[13] In these cases, another member of the lineage will eventually take over the position and the name of the *poqalla*. In the family described, this was in the process of happening. Nabulo, with a lineage of only two households, was in severe difficulties; the *poqalla* line was still in existence, but the incoming *poqalla* was a young boy. When Nabulo was as young as fifteen, he was put under extreme pressure to marry as soon as possible and to produce sons who could ensure the continuity of the lineage. Of the others, the largest lineage size was 106, and the mean was 30. In Buso village, it is possible to calculate that the *poqalla* families make up just over 3 per cent of the households in the village.

Photo 3.1 *Marrota, a regional* poqalla, *in his homestead with some of his descendants*

Photo 3.2 *The* oluahita *in Kalla regional* poqalla's *homestead*

Photo 3.3 *Bamalle regional* poqalla *and his wife (the* poqalteta*)*

Photo 3.4 *Ostrich eggs on the roofs of a* poqalla *homestead*

At first glance, the local *poqalla*s who live inside the villages appear to be quite ordinary men. They participate in work groups and meetings; they drink together with others in houses where women have brewed and are selling beer. I have attended occasions where the elders have called a man to be 'advised' about his behaviour and found later that the man with head bowed, ashamed of his misdeeds, was a *poqalla*. Little, it seems, marks them out from others on a day to day basis. Many do not wear the five iron *mugla* bracelets: some because they have not completed the formal rituals of office (this is especially the case if they inherit the position of *poqalla* from their father at an early age), others because they claim that the *mugla* are too heavy and that they interfere with digging.

On visiting *poqalla*s at home, however, it is possible to see that their homesteads are larger than usual: they often contain several large houses for people, and several large stores. There is usually an abundance of vats for brewing *chagga* and wicker baskets for storing grain. Placed in each *poqalla* homestead is an *oluahita*, a structure made from the trunks of juniper trees. Each new generation when it is blessed by the *poqalla* places a new trunk of a juniper tree in the *oluahita*. The *oluahita* thus represents the continuity of the generations (Photo 3.2).

Many homesteads include a special house which is thatched to the ground, a house where it is said the *poqalla* should sleep, as many *poqalla*s receive messages and advice in their dreams which enable them to guide and administer their lineage effectively. The houses are also often decorated with ostrich eggs, and the pots that cover the very top of the roof are elaborate and moulded in a style that could be interpreted as representing phalluses (Photo 3.4). Similarly, the very style of building of *poqalla*s' houses, like that of communal men's houses that are present in the *mora* (courtyards), is of the *kengeshia* style, that could similarly be interpreted as phallic. (In Fasha, *poqalla*s are the only people permitted to thatch the roofs in this style, see Shinohara, 1993). The phallic theme is pursued in the headpiece that a *poqalla* wears on ritual occasions, known as *xallasha*, as can be seen in Photo 3.3.

Sperber's work on the nearby Dorze people is useful here to make sense of this plethora of eggs and phalluses. The Dorze are classified by anthropologists and linguists as Omotic, like the Maale, whereas the Konso are viewed as Cushitic. However, there are certain similarities that cut across this classificatory ethnic divide. For example, Sperber describes how Dorze dignitaries called *halak'a*, like *poqalla*s, 'during rituals attendant on their entry into office adorn their foreheads with brass objects which represent, in a fairly realistic manner, an erect penis' (Sperber, 1975: 36). Certain dignitaries of other people in the region, including the Maale and the Boran (a Cushitic people), also wear this sign of office, and, in the case of the Boran, the *xallasha* is made and supplied by Konso craftspeople. Sperber writes how he was surprised how when he 'stressed this [phallic] resemblance to my informants, after having waited in vain for them to mention it to me spontaneously, my remark was taken as a joke in bad

Photo 3.5 *The eldest son of the local* poqalla, *performing a ceremony on the death of his father*

taste' (ibid.: 36). The resemblance to a penis was considered 'irrelevant' (ibid.: 44). My experience in discussions about the *poqallas' xallasha* in Konso was similar – people were reticent in commenting on the seemingly obvious. The connection, if there is one, is either unconscious or not discussed out of embarrassment.

There are other sexual associations. For instance, some regional *poqallas*, while in office, cannot cut their hair: it was said of Guufa, that no one could even touch his hair. Hair has also been regarded as a sexual symbol (Sperber, 1975). After a male animal is slaughtered in ritual the *poqalla* wears the testicles of the animal on his wrist, and later eats them. The symbolic use of eggs, phalluses and hair forms a strong theme which is central to the power of the *poqalla* and which is analysed in more detail below.

The range of *poqalla* activities

Notwithstanding the landesque capital, the Konso hold on life is tenuous: rain may fall and crops may ripen; or crops may be destroyed by pests, rain may not fall, and members of a person's family may be struck down by disease and die. This reality is represented in two contrasting scenarios in Konso thought. In the first scenario, everything prospers, the bees give plenty of honey, there is much rain, the crops ripen, the animals give birth, people are healthy and women give birth to many children. The second scenario is one of misery and ignominy: there is sickness, even madness, a lack of rain, animals do not give birth, people do not give birth; the family line suffers and may even die out. It is difficult to predict which scenario will result for a particular household, and a household has little power to influence the outcome, depending as it does on the vagaries of rain, the onset of disease, and the unpredictable process of procreation. Kahano Kalayta's words exemplify these two opposing scenarios, and his feelings about the unpredictability of nature:

> Somebody may be richer and richer. Another may become sick. It is the luck of somebody. It might be that someone has many children and that he stays well and peaceful throughout his life, for some all the children die at birth. One person digs his field and the grain ripens, one looks after livestock and they become fat, one if he puts a beehive in a tree, he gets a lot of honey, it is his luck. Another just gets poverty every year and every year. Another gets terrible problems – misery and poverty and problems for the children that he does give birth to. It is the luck of somebody.[14]

The practical, ritual and juridico-political activities of the *poqalla* all play a role in trying to mediate against these forces of nature, and to ensure that the precarious livelihoods are sustained. Each of these activities will now be examined.

Practical Roles of *Poqalla*s

The *poqalla*s lend land to others in a variety of situations. A man may borrow land to augment the holdings that he is already cultivating, and if he has access to more labour than he is currently able to use effectively. If a man who is already married marries a new wife, he needs to provide her with her own fields to cultivate. Borrowing a field may be preferable to asking a first wife to give up her cultivation of one of his existing holdings. Others may borrow land because their existing holdings are insufficient to provide them with a livelihood. As already explained, land is bequeathed unevenly to sons: the eldest usually receives the largest share, a second brother a smaller share, and the youngest brother may receive little or nothing. One option for a youngest brother is to ask his *poqalla* for help. Women, who have no recognized rights to land of their own, may also ask for land, particularly those who are most vulnerable, for example divorcees and widows. In a sample of twelve *poqalla*s, at least 13.5 per cent of the people who borrowed land from them were women. Further investigation showed that they were women who had left marriages, or who were widows. If close male kin do not provide land to cultivate, then the option of borrowing land from a *poqalla* is highly valued.

This borrowed land is important for a person's livelihood, and it influences the nature of the relationship between the *poqalla* and the borrower, as the borrower carries out labour in return for the land. Land also is important in other ways. By definition, the *poqalla* is the most senior descendant (and therefore the embodiment) of the first ancestor of a lineage. The members of each generation that is born to the *poqalla*, as they grow up, leave and set up their own home, receive a share of land on which to live. This inheritance is the same as in other households: the eldest son (in this case the new *poqalla*) receives the lion's share, but others also receive some land, described as *kotayada*. The *kotayada* continues to be divided as more sons are born. Although land may be sold or bought to change the landholdings of a household, some of the land that they cultivate will originally have come from the *poqalla*. This means that, practically and conceptually, the descendants of a *poqalla* owe an important part of their livelihood (the land they cultivate) to him, and this debt will inevitably affect the relationship between them.

Poqallas also provide other material forms of assistance. If a family is hungry they can ask the *poqalla* for grain. Just as there are strings attached to borrowing land, grain is not received entirely without consequences. It is said that the first gourd of grain is given freely, but if a person requests a second gourd of grain, then he or she must give the *poqalla* a field in return. It was difficult to find evidence of such a punitive system in operation in practice. It was clear, however, that at times of hunger and drought the price of land could drop significantly, and so, at such a time a person with resources could purchase land quite easily. The stated arrangement that a second gourd is given only in return for a field, could also be a metaphorical representation of the pressure that is brought to bear on a hungry person to sell land cheaply.

The *poqallas*' extensive landholdings mean that they have or grow resources which are needed by others. When a younger brother (*kussita*) sets out on his own and builds his first house, he needs stones, wood and grass. The *poqalla* may let him gather stones from his fields, cut grass from his grass lands, and even take some wood from the tall and straight trees that are especially cultivated in the fields for this purpose (called *oypata*). A person may also ask for stones suitable to be used as grinding stones, that a *poqalla* may have in his large fields.

If a member of the family marries and the family has no cow hide for the new couple to sleep on, then the *poqalla* will give it. At the marriage itself, the couple elope and hide for three days away from the bride's family, and they often stay in the groom's *poqalla*'s home. He will feed the couple and all their visitors with food and *chagga* mixed with butter. If a family has children, but no livestock and so no milk to give them, and no manure to put on the fields either, then the *poqalla* can give the family a cow or a sheep. The family will raise the animal as if it were their own, ultimately selling it, and sharing the profits with the *poqalla*. Interviews showed that these 'gifts' that the *poqalla* gives to his descendants and others were not uncommon. Nearly 50 per cent of one *poqalla*'s lineage said that they had received wood for building from their *poqalla*, and nearly 40 per cent had received grass for thatching. There were also cases of people receiving butter and grain, and stones for building. These are examples of forms of assistance that the *poqalla* gives to help a person overcome temporary shortages and needs. The *poqalla* is therefore a great resource for the lineage. He provides an insurance against hunger and, in certain situations, he may help someone to build a house or to marry. These forms of material assistance, some of which come with certain obligations, help with the material production and reproduction of social life.

Ritual roles of *Poqalla*s

The role that the *poqallas* play in the production and reproduction of life is represented in various rituals that they perform. One example of the ritual role is *logida* or the *arxatta ayla* ceremony that takes place at the beginning of the cultivation season. The ceremony is described by Hallpike in his ethnography (1972: 260–7). In brief, the *arxatta ayla* ceremony, literally the ceremony of 'sowing the lower part of the house compound' is the first part of the ritual. The *poqalla* sows seed on the lower part of the compound. He blesses the ground, pouring millet seed and water on the ground, and plants cuttings from a juniper tree and from a *qat* tree together with a plant called (in *Afa Xonso*) *ifiyada*. The *logida* ceremony makes up the second part of the ceremony. In the *logida*, the *poqalla* goes to all the houses of his lineage members and mixes *chagga* from his own homestead with *chagga* from theirs, and they all drink together, saying 'may the millet ripen'. Then a male sheep is slaughtered in the place where the ground was ritually sown, and the whole lineage comes and eats some of it with *dama pakaana* (boiled dough balls made from ground *pakaana*) prepared by the *poqalla*'s wives. When the *poqalla* slaughters the sheep, he blesses himself and his lineage, saying 'rain will come, cows will give birth, people will give birth, crops will ripen'. Chyme from the sheep's stomach is put on the temples of all present, and they are told 'may you not hear bad things'. The entrails of the sheep are read, and any small blood-clots on the intestines are interpreted as indications of misfortune to come, for example sickness, death or imprisonment. The patterns of the veins and blood vessels between them determine the interpretation. If such clots are present they are removed and placed on a tree, in the hope that a bird will come and eat them and by so doing take away the misfortune from the lineage.

This ritual is similar to the first fruit rituals carried out by other leaders in southern Ethiopia, for example the Maale (Donham, 1990). The ritual enactment of the planting of the crops establishes several principles that should be followed. First, it indicates that the way to plant crops is with respect and with blessings. Second it establishes that the *poqalla* should cultivate first, that he is first and most senior, and that after that, others may follow. Third, eating together and drinking *chagga* that has been mixed from all households reinforces the unity of the lineage group. Fourth, the blessing 'may you not hear bad things' protects the lineage from harm; the reading of the entrails, and the removal of any blood clots, attempts to remove any harm that has been forecast.

All *poqallas* bless people in order that they will give birth to children. Other people in the community can do so, but when a *poqalla* does the blessing it is thought to be more powerful. The most common blessings include spitting on the palms of a person and saying 'Become big/many!' (*Gutato!*) 'People become many!' (*Namma ileggi!*) and 'Give birth to a boy!' and 'Give birth to a Girl!' (*Hammiyya Hai!* or *Inenta Hai!*). In ceremonies these blessings are accompanied by being anointed with butter or splashed with milk from a cow that has just given birth.

Regional *poqallas* can also pray directly for rain. Kalla told me that people who have no rain come to him and say *'Waqa ino tiete, Waqa ino qensada!'*, which means, 'God has left us! Pray to God for us!' Then he calls to God in the following way: *'Orrariteta a unkula keraatin pisha aanato!'* which means, 'May the clouds use old [and leaky] wicker storage baskets to go and get water!', and then after they return to their villages it rains. As discussed above, it said of Kalla that when he calls to God (*Waqa*), God hears him. He is seen as a mediator between people and God, and this is one source of his power. Many *poqallas* (including Kalla) are also associated with a spirit, called *mooha*. This association is inherited with the office. The spirits appear in the dreams of the *poqallas* and give them information that allows them to guide their lineage effectively. Some of these messages may be explicit, and three local *poqallas* are credited with having powerful spirits that allow them to see into the future (Ba'ayto and Kaiya of Buso, and Sangogo of Dokato). Other *poqallas* receive vaguer messages: Onaqa, for example, told me that when he sees red sorghum in a dream he knows something bad is going to happen. Other *poqallas*, particularly the less powerful *poqallas*, have no spirit, and it is said they 'don't have dreams'. These supernatural powers are thus seen as proportional to other powers.

Some of the spirits are thought of as connecting the *poqalla* to God (*Waqa*), but this is only one way in which the supernatural associations of the *poqalla* are conceptualized and apprehended. It is one probably heavily influenced by a knowledge of and exposure to other religions such as Ethiopian Orthodox Christianity, Islam (more prominent east of Konso) and, more lately, Protestant Christianity. The spirits of the *poqalla* are also associated with the dead ancestors (*kariya*), and some spirits are connected to phenomena that display natural generative power, for example water springs. The spirit of Aija *poqalla*, for example, is associated with a nearby spring, and the spirit's name, *ella*, also means water spring.

The spirits are associated with a particular *poqalla* but they are also capricious and can act independently. For example, if a *poqalla* does not keep appropriate taboos (these become stricter as the *poqalla* become, more senior and as his spirit becomes more powerful),[15] then it is said that he can be punished by his spirit and that his spirit can even kill him. Certain *poqallas*, for example, are forbidden from going into a homestead where there is a dead body. These *poqallas* are described as *namma dawra*, which can be translated as 'people for whom things are forbidden', or sacred people. In the tightly-knit communities of Konso, attending a wake is of great social significance. It is not unusual, however, to see one or two *poqallas* who are *namma dawra*, sitting outside the compound where someone has died, excluded from paying their respects fully to the family, and also from the conversation, comforting and crying that goes on inside. If they entered the compound it is thought that their own spirits would cause them to shake uncontrollably and they could be seized with a superhuman and destructive strength. It is thought that if a *namma dawra* goes to a house where there is a dead body, he would run mad and probably end up throwing himself over a cliff to his death. In other words, his spirit would kill him. In the constraints that apply to *namma dawra*, it is possible to see that *poqallas*, the bringers of 'Life and peace', in many ways are seen as representing the antithesis of death. As much as possible, they should remain apart from it.

Judicial/political roles of *Poqallas*

As described above, the *poqallas* are thought to maintain the welfare of people by keeping the *aatta*. '*Aatta*' refers to culture, in the sense of performing the correct rituals, but also in terms of keeping an established order. This has been seen already in Bamalle's origin story, where with the drum he established order on lives that were thought to be previously wild, chaotic, and 'in the forest'.

The order is maintained symbolically, but it must also be maintained in a much more mundane sense which relates to social behaviour, and here the *poqallas* have a role too, particularly in the resolution of disputes and the promotion of peace. As others writing on Konso before have suggested, this role is most clearly evident in their role as peacemakers between warring towns (Shako, 1994; Hallpike, 1972; Amborn, 1988). In the past, when, according to oral histories, violent conflict between different villages was common, this role must have been much more important than it is today. In the case of conflict between villages, the *poqalla* would go with his *saara* (the assistants to the *poqalla*, who carry out his business for him and take messages), carrying large bells (*towna*), and walk between the areas of the protagonists urging them to make peace. Oral histories also relate how certain regional *poqallas* could also give sanctuary to an individual. If someone had committed a crime, for example, killed someone, then he could come to Bamalle's house and receive sanctuary. No matter whom he had killed, if he reached Bamalle's house, then no one could touch him. Bamalle then could help him to go to Iavello or Moyale where he would live in exile.

At a more local level, *poqallas* play a less spectacular role in resolving conflicts within the lineage. If the problem is not very serious, then the *poqalla* of the people involved makes a decision on his own, or with the help of some of the lineage elders (who are called *hiyoda*). In more serious cases, a *poqalla* may call other *poqallas* to assist, and here there are cross-clan arrangements of mutual assistance. For example, in Buso, Tigisaita and Argammayta work together to help in dispute cases, as do Sawdata and Keertita. If a dispute were to arise between people of two different clans, for example a Sawdata and an Argammayta person, then the major *poqallas* from Argammayta, Tigisaita, Sawdata and Keertita, from all over Buso would have to attend. In addition, in all adjudication, elders (*hiyoda*) must be present and their opinions must be listened to and taken into account. This serves to prevent the *poqallas* being accused of poor or biased judgement.

One case I witnessed on 6 May 1996 demonstrates this work in practice. It involved a dispute over land between three Tigisaita men. The dispute had arisen because there was a small stream between a field belonging to one person and the fields belonging to two others. One man was accused of using soil conservation practices on his side of the stream to encourage it to flow into the others' field and encroach on the others' land. The Tigisaita *poqallas* were all called together (Kaiya, Sima and Xaiyko), and also the Argammayta *poqallas* (Towaysha, Kucho, Usalle, Xelle, Turmale). Many old men were also present, and others who came to listen to the case, approximately 30 in total. The discussion was heated. Everyone who wished to give his opinion did so freely, and when it appeared that some consensus had been achieved, a decision was reached. In the end the man was fined twenty-five *birr*. Ten *birr* was given to the non-*poqallas* who had been present, the elders and other attendants, to share between themselves. They went immediately to buy *chagga*, sharing one *birr*'s worth between three people (two gourds of beer). The remaining fifteen *birr* was divided between the seven *poqallas*. Sima, as a minor *poqalla*, was not given a share. He was expected to share with Xaiyko who is considered to be his elder brother. Each *poqalla* received two *birr*, and Kucho was given an extra *birr* because the dispute was settled in the small *mora* outside the gates of his compound.[16]

Poqallas can therefore exercise judicial authority in a very practical manner, and in discussion with others, including elders. However, some *poqallas* are believed to have supernatural authority because of their spirits: Aija's spirit, for example, visits him at night and informs him of who is doing wrong. He is then able to call these people to his house and advise them to modify their behaviour. It is also believed that the spirits of some *poqallas* are capable of punishing directly those who lie when in the *poqalla*'s compound. For example, Pata *poqalla* in Dokato had such a strong spirit that when he was adjudicating in a dispute case, he would call upon those present to tell the truth or admit their fault. All knew that if they lied, Pata's spirit would become angry, and punish the liar with disease and other misfortune, destroying him and all of his family (Estiphanos, n.d.).

If the more powerful regional *poqallas* are compared to the local *poqallas*, there appears to be an increasing level of supernatural or spiritual authority of the *poqalla*, and an accompanying decrease in the mundane exercise of authority, as the *poqallas* are considered more powerful. This can be seen in the way in which regional *poqallas* rarely extract fines. Their punishment instead takes the form of curses, or more commonly of anxiety and discomfort caused by the knowledge that a person with the power to curse has been crossed.

These three roles – the practical, the ritual and the juridical – are interconnected: access to land and harvest means that a *poqalla* has the necessary resources to perform a ritual. At the ritual, those who take part also receive *chagga* and a share of the meat of the slaughtered animal. The ritual position – and the taboos that a *poqalla* must respect – lead to his being perceived as having the power to 'see' what has gone wrong and to advise. Through all these roles the *poqallas* maintain the *aatta*, the particular 'order of things', and they safeguard a person's individual existence. They also provide an explanatory framework for when things go wrong. For instance, misfortune can be explained as caused by a ritual that was not completed fully, or by an unresolved quarrel, or to some kind of disruption to the ideal order. There are similarities here with other work that has explored the relationship between ritual leaders and environmental management in Africa. Such work is not common, however, as scholars have tended to explore the areas of ritual and environmental management separately: rituals have been explored to discover the more esoteric meanings contained in them; environmental management has been explored by those interested in the urgent matters of sustaining material livelihoods. The separation between these areas safeguards the analysis from being accused of environmental determinism. But recent work suggests that ritual should not be seen only as ideational, but that it is also 'about a concrete relation with a physical world where the fertility of humans, plants and animals has to be managed in ritual as well as in day-to-day living' (Moore, 1999: 7; see also Sanders, 1998). Schoffeleers' work on 'territorial cults' identified how, in many situations, a link is made in the minds of 'cult' members between social order and environmental order, or conversely, between social disorder and environmental disturbances. The social order is both real and symbolic, and is exemplified in the strongest sense by taboos on incest and murder; but any form of 'moral laxity' is thought to result potentially in 'ecological disaster which in turn threatens the life of the community' (Schoffeleers, 1979: 5). In the context of these beliefs, ritual leaders manage the environment by performing the necessary rituals, and also by giving practical directives about what to grow, how to use and share resources, and so on. Thus in Konso, the ritual and practical are interconnected. People strive to maintain an ideal order as a way of avoiding disaster. The work and lives of the *poqallas* are central to these processes.

In Konso, the exchanges that take place between the *poqallas* and the other people in the community are economic, juridical, religious, mythological, and aesthetic. They are central to the discourse of the customary described here as Konso's orthodoxy. They structure the social organization and imbue it with meaning. As Mauss put it, 'Everything intermingles' (1990: 3). These exchanges correspond to what Mauss called 'total services', which are both the producer and the product of the moral universe in which the practitioners live (ibid.: 5). They make a connection between the person who receives and the person who gives: the *poqalla* is seen as the source of life and the protector of it; physically and conceptually, he represents a person's origins. A *poqalla* also gives the resources that are needed: he protects from 'bad things', and blesses a person to 'give birth to a boy' or 'give birth to a girl'. From a *poqalla*, life flows or, as Hallpike puts it, 'Life – health, fertility, and peace' (1972: 99). When *poqallas* receive more labour than others, it is seen as part of a tribute that is given in return for all of their work in maintaining an ordered moral world, and in trying to ensure prosperity rather than disaster. When the people give labour, it is not only an economic arrangement, but a way of honouring the position of the *poqalla* who is so intimately connected with, and significant in, their lives.

From stony to fertile grounds

A more detailed comparison with Donham's study of Maale helps to develop further the understanding and interpretation of the Konso material (Donham, 1990; 1986b). Donham's work on labour, land and social organization traces the situation in Maale from the late nineteenth century until 1975. During this time, Maale livelihoods relied mainly on slash-and-burn agriculture, and land was relatively abundant. Labour availability was the main constraint on agriculture, therefore, and it was valued at a premium. This is similar to the situation in Konso, although here, land availability is also a constraining factor on production. In Maale, as in Konso, a great deal of importance is placed on having children. Donham relates that Maale people said that a person without children died 'with a bad taste in his mouth' (Donham, 1990: 112). But this personal fertility was also bound up with a great concern for fertility in the broader sense: 'bringing of rain, the sprouting of crops, the births of children, calves and goats, the creation of all wealth in Maale and the corresponding destruction of all enemy generative power' (ibid.: 94).

The generative forces that produce fertility in Maale were unevenly distributed across a social hierarchy. At the top of this hierarchy was a ritual king, *kati*, who in certain ways resembled a *poqalla*: he carries out first fruit ceremonies; he was forbidden from cultivating his own fields; when inaugurated he wore a phallic iron headpiece that was like a *poqalla*'s *xallasha*. Below the ritual king in the social hierarchy were chiefs, below the chiefs were sub-chiefs, and the nature of the uneven relations between these leaders was reproduced in uneven relations throughout the society: the heads of maximal lineages were considered above the heads of minimal lineages, elder brothers were considered above younger brothers, and men were considered above women. The axes of inequality mirrored each other at different scales, and the inequality was manifest materially and ritually. In material terms, inequality was evident in the way in which the inheritance of property was weighted to the first born. In ritual terms, those further down the hierarchy were forbidden from consuming the produce of their fields until those directly above them had completed the first fruits rituals: hence, through ritual, elder brothers controlled the production of their younger brothers. Again there are parallels with the Konso case, where there is also a social hierarchy in which 'inequality is pyramided upon inequality' (ibid.: 120): regional *poqalla*s are considered senior to local *poqalla*s, local *poqalla*s are senior to their descendants, maximal lineages are senior to minimal lineages, elder brothers are senior to younger ones, and men are senior to women. Although most partners in these uneven pairings are related by blood or marriage, the inequality is very real, and evident in the material rights that each has to property; and also in their relative position as either givers or receivers of ritual blessings.

Donham's argument is that, in Maale minds, it is the uneven control over fertility that underpins the social hierarchy. Fertility resides in the ritual king who is the embodiment and the source of fertility. Fertility flows down the hierarchy, from the ritual king, to the chiefs, to the senior elders, to the more junior and so on. In order to understand why and how the ritual king was the source of this fertility, Donham points to the Maale understandings of conception and birth. In the Maale understanding, men and women both contribute to sexual reproduction: men contribute semen and women menstrual blood. But men are seen as active in this process; women are passive. It is the men who initiate and it is their actions that make the difference: 'It was the man's active force that produced the child; it was he who planted the seed' (ibid.: 109). 'Women were more passive and unreliable receptacles ... like the earth in which seeds were planted' (ibid.: 107). Cosmologically, fertility and maleness were linked together and in various symbolic ways, the hierarchy in Maale was constructed as being one of decreasing maleness from the ritual king down. The ritual king was the 'male principle incarnate' (ibid.: 112). Compared to the king, the chiefs were sometimes described as the king's wives, a reflection of their relative lack of maleness. But chiefs were themselves more male than the sub-chiefs, who themselves were more male than others further down the hierarchy: there was a 'continuous gradation of maleness from the ritual king to chiefs to subchiefs on down ... eldest brothers were more "male" than the younger ones' (ibid.).

The symbolic construction of ritual king as more male than others, gave him power over the generative forces that underpin the production and reproduction of Maale life. Fertility was seen as flowing down the hierarchy, and in return, tribute, particularly labour tribute, flowed up. The chiefs organized 'commoners' to work on the ritual king's fields and the chiefs received labour tribute themselves. Thus in order to understand the movement of labour in Maale, it is necessary to understand the gendered construction of the ritual and the symbolic control over fertility: 'the giving of tribute was bound up with ritual and cosmology, with notions that tribute to the chiefs and king was necessary in order to ensure biological fertility' (ibid.: 94). Donham draws on the work of Meillassoux (1981) to emphasize that control over reproduction is as important as, perhaps even more important than, the control over production in structuring society. But the key point is that this control over both production and reproduction resides not only in control over the material resources, but also in the control over the non-material ritual and symbolic resources of production.

This ritual control, Donham argues, is as effective in maintaining the hierarchy as any coercive power. It is particularly effective, as the ritual and the material are intertwined and inseparable. The inequalities are reinforced for example by 'support in cases of social conflict, help in marrying and sponsoring wedding feasts, and, perhaps most important, access to land on a basis guaranteed by kin ties' (ibid.: 119). These other forms of assistance further discourage people from resisting or rejecting the 'system' altogether.

This analysis is helpful in examining the Konso case. The Konso agriculture is produced out of a set of social relations that is far from egalitarian. The *poqallas* control a larger amount of land than others, and they lend it out in return for much needed labour on the fields; but they also receive labour because materially, politically, and ritually, the *poqallas* play a role in promoting the fertility, in the broadest sense, of their lineage and others. Materially they give food to eat when someone is hungry, and they can give sanctuary to a newly married couple during the confinement preceding their marriage ceremony. They can provide butter necessary for the blessings of the marriage ceremony and the cowhide for the couple to sleep on in the future. They can give land to cultivate, they bless people to have children, they pray for rain, and they resolve disputes. All are key components of the production and reproduction, or the fertility, of Konso life.

In Maale, Donham argues that the generative forces of fertility are thought to reside in maleness and that the ritual king at the apex of the Maale hierarchy is more male than anyone else. The *poqallas* too have symbolic qualities that may be associated with an exaggerated maleness: their phallic headpiece (*xallasha*); their consumption after ritual of the testicles of a slaughtered male sheep; the phalluses on the pots of their homes. These could be interpreted as emphasising the maleness of the *poqallas* as in Maale: fertility resides in maleness; the *poqallas* are more male than others; therefore the *poqallas* control fertility. But further investigation of the Konso case shows that this interpretation is not the full story. The actions of the *poqalla*s that have been described in this chapter – the blessings, the political patronage, the loans of land and food – are described in the Konso language not by a male metaphor, but by a female one: breastfeeding. This puts a very different perspective on the construction of fertility, on the nature of the position and the power of the *poqalla*, and also on gender relations in Konso.

When I first heard *poqalla*s described as breastfeeding others, I was surprised, but the idea is present in a text of a speech at a public meeting recorded by Hallpike in the 1960s. This reflects the institutionalized and lasting nature of the metaphor and its use for describing these processes. In the speech, one man says 'May God bless you with milk gourds… Listen… So that we may suck at the breasts of the seven priests; when you suck them you prosper' (Hallpike, 1972: 85).[17] I asked Kahano Kalayta and Kahano Kolobo why *poqalla*s are said to breastfeed the people. They replied:

> People sow their lands. When hunger comes, and when famine comes, and everyone has hunger, they go to the *poqalla* and take one container of grain. You know this *erurota* [unfermented grain drink]? At the time of the big rains, it is not in all houses, hunger comes at this time, and *poqallas* give to all people. They can dig up the *pakaana* [an edible tuber] and use them. Because *poqalla*s are strong, that's why people breastfeed from them. People take grain from them, as a loan, and after their grain has ripened they return it back. Those families who cannot return it, are tolerated by the *poqallas*, the *poqallas* don't mind and they don't demand it back. That is why [we say] they breastfeed. The *poqalla*s can give a cow also; also a sheep. Butter, if people are in need, for example in Karate when someone has died, they smear that person with butter. If they don't have butter they will get from the house of the big family, and they use it. The animal skin that is used to wrap the dead body is also taken from the *poqalla* family. People take a terrace from the *poqalla* family: He gives it to them and says 'Sow it and eat food there'. They also give honey. When you marry you go and marry in the *poqalla*'s house and stay there for three days, and all the people escorting you eat food there. The *poqalla* gives grain and *mida* [green vegetable] at the wedding. That is why it is said that the *poqallas* breastfeed others.[18]

The application of the female metaphor of breastfeeding to this process is striking. Once, when discussing the shape and meaning of the phallic *xallasha* headpiece, a man did venture a more explicit interpretation, and said that it was a breast. At the time (early in the research) I dismissed this as either delusional or as his effort to spare (or encourage) my blushes. Subsequently I realized that it could be interpreted as another reflection of the connection of the *poqalla*s with breastfeeding.

One interpretation is that this metaphor is a symbolic device for appropriating the powers of procreation which are usually thought to reside more with women than men. Such ideas and interpretations are known from the anthropological literature: in New Guinea for example, the initiation rites of the Sambia are interpreted as reflecting a deeply held fear among men that women are more powerful and reproductive than men (Herdt, 1987; Gledhill, 1994). The authority and dominance of men in Sambia is thought to depend on their masculinity, which itself is considered to reside in 'a bodily essence, a substance akin to life-force' that exists in semen (Herdt, 1987: 31). In the initiation rituals, however, certain phallic male objects are said to have been made by women, and semen (the male life force) is equated with mother's milk. For Herdt, these ritual subversions of male substances at the centre of male rites, demonstrates a deep belief among the Sambia that women are 'naturally fertile and productive' and therefore powerful, whereas 'men are not' (ibid.: 1987, 190).

In Konso, in the merging of the symbol of the *xallasha* and the breast, the substances of semen and mother's milk could be seen as being symbolically merged. However, this is not to conclude that the Konso consider female procreative powers to be more powerful than male, although the neatness, consequences, and comparisons with previous work make the interpretation seductive. In Konso, women and men are thought to contribute equally to the reproductive process, and there is some evidence of a fusion of male and female symbols. These characteristics are not exclusive to Konso and can be found in a growing body of literature on the combination of different gender categories in ritual and other more practical transformational processes (Moore *et al.*, 1999; Sanders, 1998; Herbert, 1993; Delaney, 1991; Jacobson-Widding, 1990). This literature emphasizes the transformational and generative power that is created by the fusing of symbols of different gender categories. For example, Herbert shows how in the process of transforming iron ore into refined metal, many blacksmiths in Africa use forges displaying both male and female symbols (Herbert, 1993). The bringing together of the male and female symbols makes possible and makes safe the magical transformation of formless iron ore into functional metal. In Tanzania, Sanders shows how the Ihanzu believe that rain is good and results in a harvest only if there is a combination of rain that is male and rain that is female. The ability to bring these two kinds of rain resides in ritual chiefs (male and female) who produce rain by ritually combining together other substances that are male and female. These combinations of male and female substances mirror the reproductive processes and the human life-cycles, and result in the generation of productive forces and the continuance of life.

In the *poqalla*, elements of maleness and femaleness are brought together in the fusion of phallic symbols and breastfeeding. The *poqalla* is thus different from other people, who are either male or female. According to Chasseguet-Smirgel (1983), the combination together of different categories breaks down boundaries between gender categories which are usually clearly maintained, and imputes a magical quality in which someone perceives himself or herself, and is perceived, to become closer to God – or the original powerful generative force. The ritual combination of these gender categories gives him/her control over fertility, partly from the uniting of maleness and femaleness which itself mimics the reproductive act (Bourdieu, 1977; Moore, 1999; Sanders, 1998). The normally bounded and distinct oppositional categories of male and female are fused in a process that is dangerous but that also results in generative power.

This analysis might appear quite tenuous if there were not further evidence from Konso and from the region of the crossing of gender boundaries or the merging of gender categories to create a powerful force. In Fasha, in Konso, there is another ritual leader who was inaugurated by the *poqalla*s of the area, and who is known as *Lehemaita*. (He was chosen from another category of ritual leader known as *shorogota*, discussed in the following chapter.) The *Lehemaita* is in office for nine years, and during this time he blesses the people and curses pests and crop diseases and anything else that might do harm to people or their harvest. On taking up office he is circumcised (unusual in Konso), and while in office he must wear an *oowa* – a skirt like that of a woman. He must sit down to urinate, it is said, 'like a woman', and he must not have sexual relations with a woman. He is not thought to be a woman during this time, but to be a man who does certain things 'like a woman'. Another small but significant example is the way in which the word *xaay,* which means to give birth, is applied to men in relating the oral histories of their lineages, such as in the account earlier in this chapter which describes how 'first, Kalla gave birth to…'.

Among the Gabra, the Cushitic camel herders that live among the Boran to the south-west of Konso, the most important male elders, *d'abella,* are considered to be like women (Wood, 1999). Unlike the hierarchy in Maale, as men become more senior among the Gabra, they become more female. In Wood's book, *When Men Are Women*, he discusses how the *d'abella* may still have sexual relations with women, but, like the *Lehemaita,* they must sit down to urinate. In certain situations, the Gabra also blur other oppositional categories which are considered to be gendered and either male or female, for example, north-south, inside-outside. Through this, the normally separate categories of maleness and femaleness are blurred and transgressed, and the magic, authority and position of the *d'abella* is generated.

The *poqalla*s in Konso therefore hold a magical-ritual position that represents the origin of the people, and that safeguards the continuity of their fertility through providing material resources, blessings, and maintaining peace. In breastfeeding, they mother the rest of their lineage, but breastfeeding is not only a ritual process; it also refers to material gifts of food and land which can, in certain circumstances, help a household materially to maintain its living. In breastfeeding, the ritual, the material, the juridical and the moral are 'intermingled'. The metaphor of breast-feeding captures the way in which these exchanges are 'gifts': mother's milk is something that, ideally, is given freely; it is a symbol of unselfish giving, of giving without any expectation of return. But as Mauss (1990) has demonstrated, even the most disinterested gift brings with it the obligation to return the gift, in a similar form or in another form, for example in the case of giving labour to a *poqalla*. Even a gift that is given to someone who is closely connected to the giver, as the *poqalla* gives to his descendants, creates obligations, and has the capacity to create or reinforce an inequality between the giver and the receiver. In this case, the metaphor of breastfeeding disguises the 'gifts' that are given in return (Bourdieu, 1977; Moore, 1999). Through breastfeeding, the very important labour is received as a reciprocal giving – the labour that makes it possible for the *poqalla*s to construct soil and water conservation structures.

There is, therefore, a strong resemblance between the discourses relating to people's behaviour and to the management of land: the ideas and concerns relating to maintaining the fertility of people mirror the ideas and practices which relate to maintaining the fertility of the land in Konso. The ritual combination of maleness and femaleness gives the *poqalla*s their ritual power over fertility (of people and beyond), and also justifies their position. It is this position which enables them to obtain the labour which is physically applied to maintain the fertility of the land. The cosmological and practical engagements of the Konso with their world (which is both a physical world and a world of ideas) are two sides of the same coin. The consonance that exists between the physical and the ritual engagements gives shape and legitimacy to both of them: each influences and gives strength to the other. Through their role in ensuring the interconnected ritual and practical fertility of people and the world, the *poqalla*s literally and symbolically turn the stony grounds into fertile ones.

Conclusion

Scholars seeking to understand rural livelihoods and landscapes have often viewed ritual as a mere curiosity, or as something limited to the sphere of religion and culture and of marginal relevance to the more pressing concerns of maintaining a living. This chapter has shown that ritual ideas and ritual powers are often central to maintaining that living: partly because they justify the position of those who have a role in organizing the use of resources; and partly because emic ideas about the environment and productivity are frequently formed, practised and reinforced in ritual. As Delaney's (1991) work has shown, conceptions of fertility structure wider relations of power. In the Konso orthodoxy, matters related to fertility of people and land are managed ritually. As ritual matters replete with meaning and emotion, they are extremely powerful discourses. As discourses, they should not be dismissed as 'not real'. I have not gone so far as to say that the rituals and the blessings may actually make a difference to the outcome and the fertility of the land and the people, as the people practising these rituals would contend. It is not possible to draw such a definite conclusion, and I agree with Blakely *et al.* (1994) that the safest course is to maintain a methodological agnosticism. The evidence given here, however, suggests that ritual is vital to the practical management of the environment, and to shaping ideas about that environment which have real practical consequences. The rituals may not be articulated in terms of conservation or managing the environment, but a very real connection is made between the ritual maintenance of a moral social order and environmental sustainability.

Further discussion is needed on the implications of this suggestion that Konso landscape and livelihoods are produced as much ritually as they are materially. Firstly, it is of practical importance in the current development context, in which development organizations are attempting to find grassroots institutions through which to implement the currently popular community-based development programmes, many of which have an environmental component. My experience of such a development project in Boran, neighbouring Konso, revealed that the development agents identified a series of indigenous institutions which had a role in natural resource management, from the *aba gada* (the head of the generation set) on down. At first it was thought that a partnership could be built between development organizations and these institutions to promote improved natural resource management with a greater input from local people, and building on their indigenous knowledge. Some time later, however, these efforts at partnership building broke down, and one of the reasons given was that the institutions in question were not natural resource management institutions: they were *ritual* institutions. The two were seen by the development organizations as mutually exclusive. There were likely to have been additional reasons for the breakdown in this development approach (Watson, 2003), but I hope this chapter has demonstrated that drawing a strict distinction between the ritual role of institutions like that of the *poqalla* in Konso, and their more practical engagement with the management of land and labour (or, to translate it into technical development-speak, 'their involvement in natural resource management') imposes a false dichotomy between the ritual and the practical. The two processes are not mutually exclusive; more commonly they are deeply interrelated in terms of the values of the people, and their perceptions of what they are doing and with whom they are doing it.

Second, ritual and religious change is often thought to be something that takes place separately from other aspects of life. This chapter demonstrates how ritual and religious practices are deeply implicated in practical and political matters. Ritual and religious change will have widespread repercussions. Third, when development workers talk about engaging with culture, 'taking culture seriously' and 'building on local culture' (Radcliffe, 2006; Crewe and Harrison, 1998), it is important to realize that the culture they are referring to is not a set of secular forms of organization and customary practices to which religion and ritual may be added or taken away at will. The institutions involved in the agricultural practice are imbued with wider religious and political significance. The way people relate to these institutions may be, in part at least, emotional. As a consequence, as time goes on, the institutions and their authority may be accepted or rejected for entirely legitimate reasons that have nothing to do with practical or economic matters. They may, however, have practical and economic consequences.

Finally, this chapter has explored what might look like a functionalist's dream. The *poqallas* are at the centre of a functioning system in which ritual, religion and environmental management work together. The *poqallas* provide a moral economy in which the moral dimension is provided by the ethos of Konso *aatta* (which the *poqallas* embody) and the obligations of kin and non-kin to honour their *poqalla.* The 'system' appears to operate for the public good, protecting the poorest and maintaining the landesque capital. The way in which the practical elements of *poqalla* activities are embedded in ritual and religious beliefs and practices demonstrates how difficult it would be to reconstruct similar social and cultural institutions in other environments elsewhere. Moreover, although the system 'works' for some, it is already clear that some people benefit more from this 'system' than others. Even the degree to which *poqallas* are successful at manipulating the possibilities afforded by the system varies between individual *poqallas*. The issue of differential benefits and unequal power relations is examined in more depth in the following chapter.

Notes

1 I have not altered Shako's spelling of the *poqallas*' names in this table. However, I have continued to use the spelling of the main *poqallas* that I used in the field which was the most accurate I could establish with informants and assistants.

2 The table demonstrates that the label 'clan head' could be misleading, as it would be logical to expect it to mean that each of these *poqallas* is head of one of the nine clans. As shown, there are several regional *poqallas* of each clan, (and four of Shako's clans, Kantayta, Sellata, Iktayta and Malla are all sub-clans of the Keertita clan [Shako, 1994: 81]). I refer to them in this work as regional *poqallas*, as their positions are recognized over a large area. Kala, Quufa and Pammale (henceforth Kalla, Guufa and Bamalle) are the most important regional *poqallas* in Konso. Marrota regional *poqalla* was also interviewed as part of this work.

3 Titiipa is named as a Konso clan *poqalla* but he lives in Gidole, among the D'irashe people. There the Titiipa is a big '*D'amma*' family, a D'irashe institution which shows very many parallels with the *poqalla* institution (Hansemo, 1992). This illustrates how remote some of these *poqallas* have become to Konso and/or how distinct borders between these ethnic groups (and the groups themselves) could be relatively recent phenomena. As with the other brothers that were born with Bamalle, and who stayed in Boran or Burji, it also shows how there is a great deal of cross-over between peoples who are considered now to be ethnically distinct.

4 Other stories refer to babies born in gourds carrying tools for weaving or blacksmithing;

these babies are purported to be the ancestors of *xawuda* (craftspeople).

5 This kind of outsider is also discussed by Turner (1969).

6 Tenayu, 1996.

7 Bamalle, February 1996.

8 One reading of these stories would be that, with the coming of the *poqalla*s and the *aatta* they brought with them, nature is transformed into culture. *Aatta* involves behaving socially and enabling rain to rain, cows to give birth, people to prosper and so on. Nature and culture are interwoven here, and there is no implied nature-culture dichotomy (compare to Ingold, 2000).

9 See Murphy and Bledsoe, 1997, on power and oral histories creating fictional kinship.

10 Kussia Towayta, Roba Kudana, Tefera Kamayda, 1990.

11 Although strictly it is said that a new *poqalla* should be a descendant of a second wife, there are instances of a new *poqalla* descending from a younger brother or *kussita*. In such a case it seems the genealogy of a person can be fudged, if he has amassed enough wealth and has very many descendants. This is one of the main roles of a *poqalla*: to ensure the fertility of his lineage, and through blessings to make them many, and thus a successful lineage.

12 In certain cases, his descendants still pay respect to the original *poqalla*: for example in the killing of large game animals, a lion or a leopard or a python. The skins are still presented as tribute to the first *poqalla*.

13 *'Tiga helumi, xelliteta godidi, uma damta koda ieteninjan, uma unta kosa kella inkito – maqarra!'* (Kahano Kalayta and Kahano Kolobo, 1996).

14 Kahano Kalayta, 1996.

15 The taboos that a *poqalla* must keep are most extreme in the case of the regional *poqalla*s who are forbidden from digging the ground themselves. They must also eat food prepared only from their own fields.

16 The accusers did not appear to get a share of the fine in this case, though in most cases people are asking for some compensation. In this case they were satisfied by stopping the man from continuing his aggressive land-grabbing practice and by reclaiming the rights to their land.

17 I believe the 'seven priests' refers to those *poqalla*s in Buso who preside over *mora*s (public meeting houses) outside of their homes. The ownership of these public ritual spaces denotes and indicates their senior position and power.

18 Kahano Kalayta and Kahano Kolobo, 11 December 1995.

Four

Political Life of Agriculture

In this chapter, the study of the *poqalla*s is developed further to examine more critically the nature of the power relations that exist between the *poqalla*s and others. The previous chapters have demonstrated that the *poqalla*s play a role in turning the stony grounds into fertile ones, which means that they exist as a form of institution central to the intensive agriculture. Development organizations have 'discovered' the importance of the institutional dimensions of environmental management in recent years, and indigenous institutions are seen as particularly valuable resources because they are considered to be already existing institutions that can potentially be harnessed for environment and development ends. These approaches to working with indigenous institutions tend to focus on identifying and illustrating the role that they play in environmental management. In arguing for these institutions' positive role, there is little time to think more critically about the nature of the power relations between these institutions and the others in the community. If anything, development organizations normally assume that the institutions are the legitimate representatives of the community and working for the common good.

This chapter argues that development organizations need to think more carefully about the nature of the power of the institutions with which they are working. For example, are they simply representatives of the community among whom they live and institutions that provide a coping mechanism for the poor and sustain the landesque capital? Or was the *Derg* government right when it viewed these institutions as privileged landlords, who enjoyed disproportionate levels of opportunities, access to resources and life chances? These questions are of practical and political relevance to development organizations, because, if they form partnership with indigenous organizations such as the *poqalla*s, their actions will transform the power relations on the ground. Their actions may not always improve the lives of the poorest. On the contrary, they can exacerbate or entrench existing unequal power relations.

This chapter therefore turns its attention to more bold and political questions about the nature of the *poqalla*'s position and power. Other work on indigenous knowledge, indigenous institutions, culture and development, has seldom explored the messy politics involved in the processes that are being studied. Older debates in anthropology and archaeology about agricultural intensification have examined what was referred to as social stratification, and an examination of their approaches serves as a starting point for exploring the issue. It is also helpful to examine debates about class in Africa, and to assess the usefulness of the concept of class for understanding relations in Konso. Relations of inequality do not exist in a vacuum but are the products of particular histories. In this chapter, I outline the early history of Konso and the way in which the early relations with the Ethiopian Empire impacted on the Konso society and institutions. Finally, although the key metaphor of breastfeeding has already been discussed in some depth, I explore some of the other metaphors that the Konso people use for talking about these power relations. These metaphors are used to structure and institutionalize power relations. The real value of examining the metaphors lies in the way they provide a window on how relations of inequality are thought about and experienced in Konso.

Social stratification and intensive agriculture

In work on intensive agriculture, inequality is discussed in terms of levels of social stratification. In the wake of the classic study of agricultural intensification by Wittfogel (1957), *Oriental Despotism*, the terms of the debate have focused on the extent to which social stratification and intensification are causally linked (Hakansson, 1989; Brumfiel and Earle, 1987; Widgren, 2004). Social stratification here refers to increasing degrees of status and occupation differentiation in society. It includes the emergence of elite groups with superior degrees of economic, political, social or symbolic power and wealth. Models that link social stratification and intensification assume a continuum of social organization and form of production from more egalitarian extensive livelihoods, for example, hunter-gatherers, to more complex and more unequal societies practising more intensive agriculture (Service, 1975). As societies become more complex, forms of institutions such as chiefs and big men may emerge. 'Big-men' and 'chiefs' have been interpreted as forms of political centralization that represent intermediary stages in the development of the state (ibid.). 'Big-men' are akin to entrepreneurs who rise in status because of their own efforts and skill, acquiring followers and control over them (and therefore social position) whereas 'chiefs' have position and authority and control over others because of the office into which they are born (Sahlins, 1963). Sahlins (1963) and Friedman (1975) portray the organizational structure based on chiefs as a development from the institutionalization of the big-man organization. The evidence from Polynesia suggests a developmental sequence from minor small-scale chiefs to major complex chiefdoms. However, even the work of these authors reveals that the character and roles of these different positions are not wholly exclusive: the power and position of some chiefs depends partly on their personal ability to influence people and engender their support. Maintaining the support of what Bourdieu describes as a 'clientele' (Bourdieu, 1977) is achieved through conspicuous acts of generosity, which are as important in the case of chiefs (who to some extent inherit their clientele) as they are in big men.[1] Both Bourdieu (1977) and Donham (1990) agree that these forms of patronage are highly effective in maintaining the support of people, when other sources of control, for example, coercive power, are absent or more costly. Such patron-client relations are common throughout Africa and they also feature prominently in literature on the African state (Clapham, 1982; Bayart, 1993; Chabal and Daloz, 1999). There, explanations for the personalized nature of African politics have been based on the idea that the structures of patron-client relations have been extended in the postcolonial period from the level of community to the government.

The position of *poqallas* can be interpreted according to this framework. They are like chiefs in that they inherit a clientele in terms of their descendants, but through the conspicuous acts of generosity that they perform in 'breastfeeding', they also create obligations and obtain the support of others through a deliberative process of enlargement of their faction, behaving, in other words, like big-men.

The models of social stratification and intensification have viewed the emergence of political centralization – elite groups, chiefs or big-men – from two perspectives. The first is described as 'adaptationist' because it relates the emergence of elites and forms of political centralization to the need to manage the resources and inputs needed for intensive agriculture in particular environments: '[p]owerful centralised leadership is seen as developing in environmental and demographic contexts where effective economic management is either necessary or especially beneficial' (Brumfiel and Earle, 1987: 2). Elites enable production to be organized at a level beyond that of the household: they allow land and labour to be scheduled for maximum use, and for the construction of forms of agriculture that need the coordination of many people, for example irrigation canals and terraces. Political centralization is thus seen as the product of particular environments in which the 'ability of political leaders to organise more effective subsistence economy... [through] centralised decision making, particularly in scheduling labour and land for alternative uses...[is advantageous] where aridity forces people to devise particularly complex subsistence strategies' (ibid.). Political centralization also provides a form of insurance and protection in times of hardship: elites can accumulate food in good years and distribute food in bad years. It can also be connected to trade, as it can facilitate exchange both within and across borders: much political centralization develops to manage trade outside its borders, or to diversify production and then facilitate exchange between different specialists (Sahlins, 1963). In short, the development of elites is envisaged as able to 'enhance productivity, diversify subsistence and provide insurance against food failures in one sector or another' (Brumfiel and Earle, 1987, 2).

This model is applicable to the Konso case: *poqalla*s organize labour at a level beyond that of the household, and facilitate the construction of landesque capital. They provide food to the hungry in times of need, and land to the landless, and hence serve as an insurance mechanism. Their role in trade is less well developed, although oral histories tell that regional *poqalla*s once had a control over markets. In these histories, *poqalla*s claimed a tax of salt each year from all traders who came to the market.[2] Competition for control over the market in Dokato has been a strong feature of the relationship between the regional *poqalla*s Bamalle and Kalla, and a reason for conflict between them. This suggests that there may be some truth in the claims that *poqalla*s controlled markets and received certain benefits from their control, as this would be a reason for fighting over it. Donham's work suggests that control over trade may be more developed in the Maale case, and may have been a stronger reason for the development of chiefs and kings in Maale. The ritual kings had a monopoly over the valuable trade in ivory and skins that came from the immediate hinterland of Maale (Donham, 1986b).

The usefulness of the 'adaptationist' model lies in the way in which it underlines the role of the *poqalla* in production, but it portrays the *poqalla*s as 'benevolent dictators', and the relations of inequality as completely benign, functioning largely for the good of the wider community. This model says little about the experiences of those who are situated differently in the power relations. The power relations are described as static and stable– an adaptation which enables the best use to be made out of the difficult conditions presented by a particular environment.

The second, more political explanatory model describes agricultural intensification as resulting not from the needs of production in a difficult environment, but from the opposite. Surplus is needed to fuel political competition that exists between emerging elites; this leads to an increased demand for surplus, for example in the form of tribute, and increased demands provide the impetus for intensification. In this view, the elites 'consciously and strategically employ specialization and exchange to create and maintain social inequality, strengthen political coalitions, and fund new institutions of control' (Brumfiel and Earle, 1987: 3). When individuals participate in the competition for power for their own benefit, relations between individuals and groups are experienced as more unequal.

There has been a debate between the proponents of these two views, in order to try to explain *the* reason for the emergence of social stratification and intensification; and hence ultimately to describe the reason for its emergence in one place rather than another (Service, 1975; Hakansson, 1989). The debate has led to the positions becoming essentialized and being seen as a question of environmental determinism versus a quest for power that supposedly exists in all individuals. I have shown how to some extent the Konso case fits the first model, although it cannot provide a full understanding of the relations and the experiences of the people. In what follows, I show that the Konso case can also be seen as fitting the second model, and how this has implications for understanding the nature and experience of power and social inequality in Konso.

Clues from 'breastfeeding' and burial statues

Questions about the nature of power are difficult to examine directly, through interviews for example, as questions may be seen as overly leading and subjective. But conclusions can be drawn from carrying out an 'accountancy' of the exchanges of services that take place in 'breastfeeding'.[3] Such an accountancy of exchanges between people is not easy, as the exchanges are often subtle and varied, and may be spread out over a long period of time. But clues about the exchanges can be found by analysing the wooden burial statues, known as *wakas*, that are erected following the burial of important and wealthy men. The *wakas* are a form of material culture in which elements of the process of 'breastfeeding' are represented. They can be analysed to show the process of 'breastfeeding' in more detail, and to reveal something about the nature of power in Konso and the way in which it is thought about. The *wakas* go back over several generations, and they can therefore be used to explore the similarities and differences in the nature of power and the exchanges that have taken place over time. Many of the older *wakas* have been stolen or are now falling into disrepair, but there is enough evidence to suggest a degree of continuity over time in the process of breastfeeding and relations between people.

Wakas are constructed after a person's death, and ostensibly do no more than commemorate and celebrate their life, particularly if they were successful, important and wealthy. As *poqallas* are important and wealthy men, the *wakas* often (or even usually) represent the life of a *poqalla*. *Wakas* are placed in prominent places, for example the gate of a village, in a *mora* (public meeting courtyard) or by well used paths to a village. They are very expensive to make and not everyone can meet the costs of making one. Several mature juniper trees must be cut to make the basic figures, and the carving of the statues can take several months. During this time the specialist craftsman lives with the family for whom he is working. In addition to the payment for the work, he must be supplied with good food including meat, honey and alcohol during his stay.

The *wakas* are a kind of three-dimensional obituary. The *poqalla* (or whoever is commemorated) is the central figure, complete with prominent

Photo 4.1 *Older* waka *with two wives. Two figures represent those killed by the* poqalla*'s lineage. Stones hidden by vegetation*

Photo 4.2 *Relatively new* waka. *Stones are at the front, fixed in concrete*

xallasha (phallic headpiece). He is flanked by his wives, whose place of origin can be seen in the kinds of hairstyles that have been carved. Beside the wives are figures of people that he killed during his lifetime, or, more precisely, those his lineage killed during the time when he was *poqalla*, as a *poqalla* is forbidden by taboo from killing another person while in office. At the front of the main statues lie wooden sculptures of the game the *poqalla* killed, or that his lineage killed and presented to him, for example leopards, lions, or baboons (see Photos 4.1 and 4.2). Before the statues also lie stones which indicate, not, as Hallpike (1972) suggested, the number of fields owned by the man who is commemorated, but the number of fields that he acquired during his lifetime, by any means. Significant to this research is the fact that the stones are divided into two kinds: those that are 'standing' and those that are 'sleeping'. The flat, 'sleeping' stones represent the land bought or received from the lineage; and the standing stones show the land bought or received from non-lineage people.

The way in which the number of fields that a *poqalla* acquired in his lifetime is represented, as opposed to only the number of fields that he owned, is significant. The way in which each new field is represented alongside the number of wives that he married, and the number of kills that he made, reflects not only the importance of *having* land today and in the past, but the importance of *acquiring* land. It suggests that for some time, a pride has been taken in the open pursuit or even struggle to acquire land. It is comparable to the hunting of prestigious animals, the killing of enemies and the marrying of wives.

Land is acquired through the successful operation of 'breastfeeding'. Through 'breastfeeding', the support and allegiance of those who receive the benefits are ensured, but this support and allegiance are also manifested in much needed labour. The labour is then converted into other resources: through receiving labour the *poqalla* can cultivate more land, obtain more harvest, and ultimately, he may be able to convert this harvest into more land. For example, since the monetization of the economy, harvest may be converted into cash, and land may subsequently be bought. Even before the monetization of the economy, the *poqalla*'s significantly greater wealth in harvested produce could be converted into landholdings. This possibility is demonstrated in the indigenous food relief practices in which a first gourd of grain is given free, but the second is given for a piece of land. It is likely that other debts to the *poqalla* could be called in similarly, augmenting the wealth in land of the *poqalla*. When land is acquired, another stone will be able to be added to the *waka* after the *poqalla* has died. Breastfeeding is therefore not only a process of distribution, it is also a process through which the precious resource of land is acquired and accumulated; the *waka*s suggest that this has been the case for a significant period of time in the past, as well as today.

The *waka*s also reflect the way in which what Bourdieu refers to as 'symbolic capital' is acquired through breastfeeding. 'Symbolic capital', is the 'form of prestige and renown attached to a family name' (Bourdieu,

1977: 179). Through giving generous 'gifts', the *poqalla* demonstrates his success, powers and abilities. Through hosting feasts and carrying out rituals, the *poqalla*'s name and reputation as a powerful leader are strengthened.

For Bourdieu, the accumulation of symbolic capital is the most valuable form of accumulation, as a good name and reputation have the self-reinforcing quality of naturalizing an individual's claims to power. Bourdieu's work on Kabylia (Algeria) also shows how symbolic capital is valuable because it can be used to acquire credit, or, and most importantly in the Kabylia case, to obtain access to the labour of others. It is worth reviewing Bourdieu's own words:

> The strategy of accumulating a capital of honour and prestige, which produces clients as much as they produce it, provides the optimal solution to the problem the group would face if it had to maintain continuously… the whole… workforce it needs during the labour period… Symbolic capital… which is readily convertible back into economic capital, is perhaps the most valuable form of accumulation in a society in which the severity of the climate (the major work – ploughing and harvesting – having to be done in a short space of time) and the limited technical resources (harvesting is done with a sickle) demand collective labour. (Bourdieu, 1977: 179; original emphasis)

The climate of Kabylia is one in which the cultivation season is short, and the demands on labour are high. Obtaining access to labour in a particular period is important to ensure successful cultivation. This is clearly relevant to the Konso situation where it is necessary to obtain access to labour because of the intensity of the agriculture.

In Konso, the *poqallas* accumulate symbolic capital in terms of reputation and prestige. They also inherit symbolic capital through their ritual offices and ritual powers, and they invest in new symbolic capital through the construction of *wakas* that celebrate the lives of their fathers and grandfathers. This symbolic capital is converted into labour as people work for their *poqalla* in order to honour him, because he hosts feasts and performs rituals, and because he blesses them, ensuring that they will live long lives and have many children.

Symbolic capital is thus important in the Konso case, but it goes hand in hand with the accumulation of economic capital in terms of land. The *poqallas* engage in accumulating both economic capital (land) and symbolic capital. These forms of capital are both used – through breastfeeding – to obtain access to the labour of others. Owing to the high labour intensity of the agriculture, access to this labour is a crucial part of an individual's successful cultivation of land, and poor availability of labour can limit an individual's ability to acquire wealth. This labour can be seen as another form of capital, which is described in the development literature as human capital. But human capital is very difficult to accumulate. The celebration in Konso of giving birth to lots of children (or in the *poqalla*'s case having a large number of descendants) can be seen as one consequence of the importance of obtaining access to labour. Having many descendants may be one way in which people can accumulate the rights to the labour of others, as descendants may labour for those they are 'born from' in order to honour him or her. But this way of acquiring access to labour is limited by biology, and is also unpredictable. It is difficult to accumulate other kinds of rights to the labour of others. Extreme forms of such rights, for example, would take the form of indentured labour or slavery, both of which require huge resources of forceful power. Access to others' labour is gained through social capital, the networks of obligation, friendship and trust, discussed in Chapter Two, and which is cemented by meanings discussed in Chapter Three. Social capital and human capital are difficult to accumulate. Symbolic capital and economic capital, which are more easily accumulated, are used to obtain access to social and then human capital, the precious labour. Acquiring capital in a symbolic form also gives its accumulation an important degree of social legitimacy (Bourdieu, 1977; Rowlands, 1987; Gledhill, 1994).

The term capital is a loaded one. Its use has been criticized in development and environmental studies because, like another loaded term discussed shortly, 'class', it brings with it sets of assumptions and values from one political, social and cultural context and applies them to another in which they are not necessarily appropriate (Fine, 1999; Jacobs, 1995). The idea of 'capital' implies that it is something that can, or should, be exchanged with other forms of capital in the pursuit of accumulation and profit. This problem is easily illustrated with reference to the analysis above. The use of the term capital suggests that *poqalla*s are only engaging in ritual practice as a way to accumulate symbolic capital and as part of a set of exchanges oriented to creating obligation in kin and others. The capital framework is in danger of failing to appreciate aspects of meaning and experience which are also involved in these practices and processes. The term 'capital' tends to imply that the individuals involved are strategically engaging in the exchanges of different forms of capital in order to maximize their position. When some of these exchanges involve gifts or help given to friends, relatives and neighbours, or participation in feasts or rituals, the notion of capital seems dangerously reductive and ignores many of the richest dimensions of people's lives.

Bearing in mind the reservations about the use of the term capital, its limited use in the Konso case is appropriate because the *waka*s suggest that there *is* a drive to accumulate land and prestige in Konso. The *waka*s make solid and legitimatize a certain version of family history in which it is legitimate to pursue and to accumulate wealth. Secondly, the term 'capital' is useful for identifying the way symbolic capital and economic capital are both involved in the acquisition of the precious resource, labour, which is more difficult to accumulate. Labour, or human capital, is in turn used to gain access to more economic capital (harvest and land) and symbolic capital (prestige and ritual authority). Together with the vital social capital, these forms of capital are, if not exchangeable, mutually connected. Access to one leads to enhanced access to another. It is through the manipulation of access to these forms of capital that the *poqallas* are able to improve their power, position and prestige. Through accumulating economic, symbolic and human capital, the *poqallas* engage in a competition for power.

The *poqallas* are in an advantageous position and it may appear that they cannot fail to accumulate wealth and power. But the process is in practice precarious, and the exchanges between *poqallas* and others are not fixed and stable. The degree to which a *poqalla* succeeds depends on his skill and luck in manipulating the process of breastfeeding. *Poqallas* do not just accumulate forms of capital through breastfeeding; they can lose capital too. The demands of kin and others for assistance, and the investments made in cultivating a 'clientele', can exceed the returns that are received in labour. Thus, through breastfeeding, the fortunes of a *poqalla* line can wax and wane, sometimes quite dramatically over generations. When discussing family histories with a *poqalla*, it is common for a particular ancestor to be picked out and described as tough (*igococki*), inferring that he was and is highly respected. Success, in Konso minds, is defined as a combination of a large amount of land, strong ritual powers and a large number of descendants and other 'followers'. This success is encompassed by the Konso word *qeydada*, which is explained as 'when somebody becomes big, then people come together and cooperate with a family' (Kahano Kalayta, 1996).

In contrast, it is equally possible for a *poqalla* to fail to accumulate wealth and people. Such a family, as mentioned in the previous chapter, is described as *helumi*, referring to the way in which it is becoming so small it will even die out. Many of the *poqalla* families interviewed for this work had fallen on bad times, and it is clear that they had not succeeded in acquiring wealth and people, and had sold land and lost their symbolic positions. The relations in Konso are therefore unstable: a virtuous circle (for a *poqalla*) can be created by accumulating land and symbolic capital, leading to increasing levels of access to the labour of a clientele; equally easily, it is possible for a vicious circle to occur, and the clientele's demands for services to outweigh the amount of capital that is accumulated.[4] Again, the *wakas* can be cited as evidence that the Konso people are aware of the precariousness of these power relations. Those who are commemorated are the people who have succeeded in a difficult process, and their descendants need to boast about it.

The *poqallas*' power and position is thought of in Konso minds as fixed and given by birth. Their position is naturalized by their claims to firstness, their origin myths, and the symbols that accompany their position, as discussed in the previous chapter. But this analysis shows that the successful *poqallas* are those who have forms of economic, symbolic, social and human capital. It also shows that the power relations between *poqallas* and others are much more dynamic than as first appears. This is also supported by an exploration of other positions of power in Konso. In the following section I review some of these positions, and show how they support this latter interpretation of *poqallas* as not only the remnants of original founding lineages, but as people who have succeeded in accumulating different forms of power and prestige. This section again shows the importance of symbolic capital and the way it is used in conjunction with other forms of capital.

Brief comparison with other positions

There are other positions of political and ritual power in Konso, which demonstrate that the picture so far, that of the *poqallas* as leaders in a sea of other people, is not the whole story. These other positions might be seen as having the potential to provide a counterweight to the authority and power of the *poqallas*. They include the *apa timba*, *xela*, *apa para*, and *shorogota*. The *apa timba* (in some villages known as the *apa paleta*) is the most important of these other positions. His position means literally the 'father of the drum', and if anyone went to Konso and asked 'who at the village level is the main authority?' the answer would not be the *poqalla*, it would be the *apa timba*. The *apa timba* is responsible for resolving disputes and maintaining order in the whole village; if there is a disagreement that cannot be resolved among the clan (in consultation with the *poqalla*), or by others, it is referred to the *apa timba*. If the *apa timba* considers a case, his fines may be higher than those at a lower level. The position of *apa timba* is held by a number of men, who take the office in a fixed rotating cycle of 36 years. Each *apa timba* holds the drum for between one and three years, and then passes it to the next person in the cycle. As the position passes from one man to the next, the sacred drum, *aibo*, goes with it, and the person holding the position is said 'to hold the drum' and must keep various taboos during the time he is in office.

Supporting the *apa timba*, and carrying out his orders, are a group of young men, called *xela*. The *xela* are in office for a period of nine years. They are headed by three men, whose positions are named after the objects that each has to carry when holding office. At the head is the *apa kuma* who carries sticks (*kummaiya*), and which literally means, 'father of the sticks', then the *apa pilowa*, who carries a knife and is literally 'father of the knife', and then the *apa lisha*, who carries a whip made from hippo hide and is literally 'father of the whip'. The names and symbols of their offices indicate the nature of their authority and power. The head of the *xela* does not inherit his office (in the way the other positions described here do) but is chosen. In the past, he was chosen by the one of the *poqallas* who has special powers to see into the future (Kaiya or Ba'ayto).

Ritual positions other than the *poqalla* include the *apa para* and *shorogota*. The *apa para* is a cyclical position like that of the *apa timba*, and there is a fixed rotating order of men who hold this office. The *apa para* means literally the 'father of the year' and it is a sacred position concerned with ensuring that there will be rain. The possession of the year is symbolized by two horns, an oryx horn (*ilikaita*), another horn for blowing and a *sakuta* feather. When there is no rain, the *apa para* may be called upon to pray for rain.

Shorogota is a fairly loosely and widely applied term throughout Konso and its meaning varies quite dramatically in different areas. In certain areas of Karate, it means simply 'rich man'. In Fasha, it is the name given to the sacred men who take part in rituals and bless the people in village ceremonies.[5] In Buso, it is quite a narrow term and is applied as in Fasha. It refers to those people who take part in the *xoxayta* ceremony, in which pests are cursed, and all 'bad things' are said to be sent far away.[6] In this way *shorogota* are said to protect the village and the crops. They also slaughter an ox in the ceremony of the change-over of the generation-grades. Some *shorogota* are called to a newly constructed house, where they bring blessings by lighting the first fire in the house with special sticks (*esaytota*).

The *shorogota*, the *apa timba* and the *apa para* are all positions that are inherited from eldest son to eldest son, like the position of the *poqalla*. The leaders of the *xela* are young men chosen because they are believed to have the qualities to carry out their responsibilities. Details about the nature of these positions are given elsewhere,[7] but it is of relevance to note here that, far from representing a set of institutions that exist as a check on the power of the *poqalla*, a large proportion of the individuals who held these offices also happened to be *poqalla*s.

In other words, there is a good deal of concentration of these offices in the hands of a few people. This is possible because, contrary to my expectations, these 'customary' inherited offices, many of which have a sacred dimension, can be purchased. Even before money was commonplace in Konso, it is likely they could be acquired in the same way that land could be acquired, in return for help, assistance or some other form of exchange. Several cases of the position of *apa timba* being purchased, quite legitimately in the eyes of the community, were remembered: in distant memory, the family of Laho had sold his right to hold the drum to Kaiya; Maxo (a family now on the verge of extinction) had also sold his right to the drum to Ba'ayto (who was three years in office; no doubt another sign of having purchased the right several times); in about 1960, Solomo sold his right to the drum for fifty *birr*. The positions of *apa para* and *shorogota* can also be bought, but the *apa para* and *apa timba* are considered more prestigious, and sell for more money.

The concentration of particular people in these offices, and the way in which they can quite legitimately be bought, points to two conclusions: first, that it is through acquiring these offices that *poqalla*s diversify and strengthen their position and acquire new political and symbolic power. Second, as anyone can purchase most of these offices, they can be seen as intermediate forms of accumulation of economic and symbolic capital. All of these offices are characterized by symbolic importance, as they confer prestige, and the office-holders must respect certain taboos, not unlike the *poqalla*s. A study of four people who held the offices of *shorogota*, *apa timba* and *apa para*, also showed that they were richer than other Konso in the land that they cultivated, but they were not as rich as the most successful *poqalla*s. It may even be possible for those with the economic resources to invest in these forms of symbolic capital as a way for a person to make himself more suitable to be considered as a new *poqalla* (see Chapter Three). Tellekaiya, for example, who is in the process of being inaugurated as a new minor *poqalla* (a descendant of Kuch'o) was *apa para* for a long time before he was considered a possible new *poqalla*. In a second example, the direct line of descent of Xombollo *poqalla* has died out, and the clan elders have selected a man from the lineage who will, in time, become the new *poqalla*. The person they have selected is Wotaro Kudade, who is already *apa timba* and *apa para* and also a member of the *xela* leadership. This man has made himself more suitable for the office because of the other forms of symbolic capital that he and his forefathers have acquired. In addition, he has shown a capacity for leadership through his role in the *xela*. In the face of this, the specification that a new *poqalla* must be the most senior descendant after five generations of the first son of the second wife of the major *poqalla*, seems to have been overlooked, a situation that is perhaps not unusual.

Power in Konso is therefore increasingly centralized, with increasing levels of accumulation of symbolic and economic capital. This symbolic and economic capital is used to gain access to the political support of a clientele, and to gain access to the form of capital that is most difficult to accumulate, labour. Those with the most economic and symbolic capital are the regional *poqalla*s, then the local *poqalla*s, then those with other positions, and finally other people. Individuals may move up and down this hierarchy in a process which may take generations.

This analysis also shows that elements of both of the models of intensive agriculture and social stratification may be present: the *poqalla*s are institutions which serve various 'adaptationist' functions, but the institutions are also produced and reproduced over time through competition for access to labour necessary for the labour-intensive agriculture. Although this latter model shows that there are unequal power relations and that they may be integral to the agriculture, it says little about the way in which the inequality of power is experienced or about the degree of agency that individuals have in the process. These models are also rather static and reveal little about the way in which these power relations have changed over time. In the next section I address some of these issues.

Power, class and Imperial history

Power is a notoriously difficult subject to examine. Linares describes power as one of the 'fuzziest' concepts (1992, 23); Mann describes the actions of power as 'messy' (Mann, 1986); Foucault admitted to Dreyfus and Rabinow that despite the centrality of the concept of power in his work, the concept 'remains elusive but important' (Dreyfus and Rabinow, 1986: ix). Although difficult to describe and understand, the nature and experience of power relations have real consequences for people's lives.

In relation to Africa, and to the Konso material discussed above, the question is: to what extent do the *poqalla*s act to serve the community – like Scott's (1976) landlords in the 'moral economy of the peasant' – or to what extent do they act to serve their own interests? This rather stark and political question might usually, and some would argue, sensibly, be avoided as over-simplistic, but it is of relevance to development organizations who wish to form partnerships with indigenous institutions to achieve particular ends. This question also emerges from a long history of work on power relations in Africa, in which there are two 'ideal types' of power relations. These have been summarized by Rowlands. In the first kind, 'power is subordinate to the hierarchy, and hierarchy is a manifestation of the sacred' (Rowlands, 1987: 52). This kind of power might be associated with the Indian caste system described by Dumont (1980), or, in Africa, with the corporate lineage or kin group that was discussed by anthropologists in the 1940s and 1950s. In the corporate lineage, the individual had no legal or political status except as a member of a lineage, and the interest of the individual was subordinate to the interests of the group (Fortes, 1953). If the lineage head had more property than others in the lineage, he was only acting as trustee of the property which was actually 'owned' by the whole group. In this form of organization, power and stewardship responsibilities went hand in hand, rather than power and individual privilege.

Another view of power suggests that the leader has personal power that he uses to his own advantage. His ownership and control over productive resources is not just in name, but is real, and sets him apart from others in the community: 'the subordinated power is in reality founded on a control over resources' (Rowlands, 1987, 52). These two perspectives posit that there is a qualitative difference in the nature and experience of the different kinds of power relations that can exist between elites and others: in the first, the relations are more cooperative and serve mutual advantages; in the second the relations are more extractive and can be said to be more 'class-like'. The difference between these two kinds of power hinges partly on the group's relations to resources; in the more class-like relations the rights to property are more exclusive, although, as will be discussed later, there are more dimensions to 'class' than merely the economic.

These two ideal types of power relations are useful for thinking through the Konso case, and they are also evident in Donham's analysis of power relations in Maale. Donham's work describes how one form of power relations gave way to the other, as the result of its history of incorporation into a wider global political economy. This idea of one form of power relations giving way to more class-like relations through the course of history is a common conception of power and history, and is a strong theme in Marx's work (see Godelier, 1986). In Maale, the positions of elites became more class-like as, over the twentieth century, the Maale became more integrated into a global political economy and the elites strengthened their control over resources. It is worth reviewing this argument, because, by exploring the similarities and differences from the Konso case, a better understanding can be reached of the nature of unequal power relations and way that they are experienced. A discussion of this work also provides the opportunity to review the relevant history of Ethiopia and the area in question more generally. The changes referred to here are from 1890 to 1974.

The Imperial creation of landlords in Ethiopia

According to Donham, Maale elites took advantage of the opportunities provided to them when Maale was incorporated into the Ethiopian Empire under the rule of Menelik II in the 1890s. Although this process was not a European colonization, it was a colonization of southern peoples by those of the north (Donham and James, 1986), and some of the methods employed by the northern colonizers were similar. Most particularly, the northerners used local customary authorities as intermediaries between the people and the state. This practice has been described for European colonies by Ranger (1983), Colson (1971), Iliffe (1979) and Mamdani (1996), who have detailed the way in which such a process transformed the position of indigenous leaders and the power relations in these societies. The work of Donham (1986a; 1986b), Markakis (1974), Clapham (1988) and Bahru (1991), shows that the colonization process in Ethiopia was no exception to this pattern.

In the 1890s, the majority of the southern regions, including Maale and Konso, were incorporated into the Ethiopian Empire by armies from the north, and the first taxation – in the form of tribute collection – was introduced into the region. The Empire at the time was divided into Provinces, and each Province was ruled by a Governor, usually somebody who had acquitted himself well during the battle in which the area had been brought under the control of the northerners. Each Governor had quite a high degree of autonomy at this time, and the state was highly decentralized. But the Governor had to collect enough tribute to supply a fixed amount to the Emperor every year, and enough to support his followers, workers and army (Bahru, 1991, 87). In order to administer this collection, the Governor subdivided the territory among his main lieutenants, who in turn assigned portions to officers under their command, and thus a hierarchy of control was established, each layer collecting tribute from the people and taking their own share (Markakis, 1974, 106). These tribute collectors were known as *melkagna*, a word best translated as 'lord', which comes from the Amharic verb *meleke*, to vanquish, to govern, or expropriate. This meaning captures the forceful and extractive nature of the process (Donham, 1986a).

The right to rule and collect tribute was called a '*gult*'. *Gults* were awarded by the crown to those serving the Empire. The nature of the *gult* holder is best summarized by Tadessa Tamrat:

> [An owner of a *gult*] was basically different from a landlord in that he had originally no right to the land itself, but only on the peasants living and working on it. The ownership of the land remained in the hands of the local people, who passed it on to their descendants according to their own customary laws (Donham, 1986a: 9).

Those dominated by the *gult* holders were called *gabbars or gabbar-melkagna*[8], a word that still has strong derogatory connotations. *Gabbar* literally means tribute giver, and tributes consisted of giving their *gult* lords part of their crops, and *corvée* labour, and also giving presents to their *gult* lords on holidays or at the time of special events, for example on the birth to a *gult* lord of a son (Donham, 1986a; Markakis, 1974). Markakis describes the tribute as gifts of honey, meat and dried grass as well as labour on the fields, grinding, transportation, building, animal care and serving as messengers and escorts (Markakis, 1974: 114). The labour given by a *gabbar* took up approximately one third of the *gabbar's* labour time (Bahru, 1991: 87) and the tribute also exacted would have been a heavy load. There are cases where those who could not fulfil the tributes demanded by their new masters were forced to give children to their rulers instead (Donham, 1986a: 40).

One example from Konso shows how the practice of taking children was not unusual. A man named Kembo explained to me how, when he was about 15 years old, his family had been unable to pay taxes and as a consequence he had been made a 'slave' (he used the word *ashkara*) to a northerner known as 'Tinfo' in the 1930s:

> He [Tinfo] put me in chains for three months. In the daytime he released my chains and at night I was tied again. To this day I have problems with my legs because of those chains. In the daytime I carried his gun for him and his *sinara* [cartridge belt] to other parts of Konso. In Gawwada [some distance from Konso] we went to find leaves for making *tej* [honey wine] and *arake* [liquor] and then I had to carry this on my head, the gun on my shoulder and the bullets across my chest.[9]

He also described the nature of the relationship that developed between himself and his captor. His captor wanted him to stay with him, and said 'why don't you stay with me and let me be your father?' But, much to his relief, his grandmother paid a fine and he was returned to his family.

Land was also appropriated by the state at this time, as in many areas a new land tenure system was implemented. The degree of variation in implementation and complexity of the changes in landholdings was so extreme that it 'practically defies coherent description' (Markakis, 1974: 129). However, the main elements of the changes can be summarized as follows: the land was divided into two categories, measured and unmeasured land. The measured land was called, in Amharic, *kudad* (the name of the rope used to measure the land) or *gasha*. This was divided up according to the *sisso* (meaning one-third) principle. One-third of the measured land was to be used by the local inhabitants, and two-thirds by the state's representatives. The measured land was generally the best land, and the state employees used it to grow food and fodder and to keep their animals on it. They worked the land with *gabbar* labour, and also had *gabbar* families working on their own fields on this land. It is difficult to ascertain where claims to taxation, *corvée* labour or taking a share of a crop ended, and ownership began, but there was a difference between what happened in the northern regions of the country, and what happened in the south. Markakis writes:

> [The northern *gabbars* maintained rights over land;] in the southern provinces, however, the peasantry cultivating land expropriated by the state lost whatever rights it had enjoyed over such land... Although the southern peasantry remained on the land ... its relationship to the land became extremely ambiguous and its rights over it highly insecure (Markakis, 1974: 112).

The unmeasured land was called *kutre gabbar*. These were areas where it was assumed that there were no permanent cultivators, such as pastoralist areas, and poor quality marginal areas, and the land was simply declared the property of the state.

The new rulers did not usually have a command of the local language or an ability to penetrate deeply into the community, so they relied heavily on the existing customary authorities as intermediaries. They called these customary authorities *balabbats*, a term which literally means 'one who has a father', and in southern Ethiopia they were generally people of notable birth who had significant rights over land (Donham, 1986a, 36). Initially, *balabbat* was an informal term to refer to any southern customary authority, but later it took on more of a formal usage to refer to those leaders who served in this way as intermediaries. Customary authorities thus became a rung in the hierarchical ladder of decentralized rule stretching from Addis Ababa to the marginal rural areas: Emperor – Governor – *Melkagna* – *Balabbat.*

In return for their service as *balabbats*, many customary authorities received privileges, which included their own *gult* grants, and some exemption from, or reduction in, taxation. They also obtained amounts of measured land, either as a reward for their service, as a by-product of their position, or by buying it cheaply (Markakis, 1974; Donham, 1986a; 1986b). Markakis describes:

> The *balabbats* were given Ethiopian titles, land and other privileges which distinguished their position from that of the ordinary peasant in the south.

> Their authority over the peasantry was supported by the northern officials, and since traditional restraints were no longer effective, this authority was often abused. The change in land holding system of the southern provinces imposed by the Ethiopian regime enabled those wielding authority under this regime to appropriate land as well as tribute and labour from the peasantry. The southern *balabbat* profited from this opportunity probably as much as the northern officials and soldiers who sought their fortunes in the new provinces. Thus under Ethiopian rule, southern societies evolved indigenous privileged groups which flourished under the protection of Ethiopian power. Ethiopian rule in the south reinforced the pattern of stratification where it existed, and introduced it where it did not (Markakis, 1974, 107–8).

According to Donham, prior to the northern colonization in Maale, the chiefs and kings existed in a relationship to others, similar to Rowlands' classification of a sacred hierarchy. Before the twentieth century, kings and chiefs had access to tribute in labour and gifts from their lineage and others who were resident on land that was considered to be theirs. The kings and chiefs were sacred, and the embodiment of fertility of their lineage. In return for their tribute, commoners received fertility in blessings from the king and chiefs; as tribute flowed up the hierarchy from commoners, to chiefs, to the kings, fertility was perceived as flowing down. The Maale kings and chiefs, however, had no 'real' control over the means of production, prior to their incorporation in the Empire:

> When the Maale gave cattle as tribute to the ritual king, they did not do so because the king 'controlled' resources like land. It is true that the king was said to 'own' the land, but in this case ownership referred to the right to collect tribute, not to control land as a productive factor (Donham, 1990, 63).

Through their position as *balabbats*, the Maale elites formed a social, political and economic allegiance with the northern conquerors (Donham, 1986b). They gained access to large amounts of land and *gabbar* labour. They also separated themselves socially from the rest of the community, for example by the adoption of a new religion and different non-indigenous practices. In this way, Maale kings transformed themselves into new elites, with economic privileges and different lifestyles from other Maale. Although, before this time, Maale kings had had a monopoly on certain kinds of trade and received tribute, their economic privilege was now transformed and made more extreme. Control over land now meant that they could claim rent from those who used it, not only from people who might cultivate it, but also from people who used it for grazing livestock, or from those who kept beehives there. The local elites controlled their customarily held land and sometimes received additional land as a reward. The local elites and northern settlers formed a social and economic alliance to their mutual advantage. Donham describes how this led to the transformation of customary authorities in Maale from ritual leaders to Ethiopian landlords:

> Over the twentieth century, these two groups – on the one hand the northern settlers and on the other the Maale king and the richest chiefs – developed into a fairly distinct class vis-à-vis surrounding Maale peasants. Northerners, the king and the richest chiefs lived in the only tin-roofed houses in Maale; they began to send their children to school years before ordinary Maale; they were adept at dealing with local courts; and they could afford [because of the tribute and land they received in return for their services] for example, to fly to Addis Ababa to consult a doctor. In sum, their style of life was different, and even the Maale members of this group, the king and some of the chiefs, were beginning to assimilate themselves to national Ethiopian culture (Donham, 1986b: 76).

He describes this process as the development of a 'distinct class', implying perhaps that, previous to this, the customary authorities did not exist as a class in Maale. Although the above quotation emphasizes the cultural and behavioural aspects of such class formation, it is clearly based in and enabled by economic developments, in particular the increasingly exclusive rights to land that they were able to exercise.

The northern rule from 1890 to 1974 was broken by a period of rule by Italy from 1936 to 1941. Markakis (1974) and Donham (1986a) agree that on the ground this meant few changes in southern Ethiopia. The Italians simply took over the network of intermediaries that had been set up. In theory, they abandoned the system of *gabbar* labour, but in Konso older informants still remember that if an Italian needed to carry goods a long distance, he would simply order somebody to do it. As one man put it 'there was no choice and no payment.'[10] When Haile Selassie returned after Italian rule, he started a programme of modernization in the country, and introduced increasing levels of taxation. But he too used the intermediaries that had been employed by previous regimes. Thus, although there were many differences over these seven decades, they all relied on the cooperation of the indigenous leaders as *balabbats*. According to Markakis and Donham, this led to three changes that made the relations between elites and others more class-like. These can be summarized as: first, an increased social difference between the elites and other people; second, increased landholdings of elites; and third, a transformation in the nature of landholdings to become more exclusive.

The Imperial creation of landlords and class in Konso

In Konso, most, if not all, regional *poqallas* who operated at a level above that of the village became *balabbats*. In each village, a local *poqalla* was appointed to act as *balabbat* for that village. The extent to which the positions of these *poqallas* changed and followed the pattern described above, varied between these two levels of *poqalla*.

The regional *poqallas*, for example, Kalla and Bamalle, gained in importance and prestige through their role as *balabbats*. Some adopted aspects of the lifestyle and culture of the northerners, and received prestigious titles of Imperial rank in recognition of their service and importance. The title *Qagnazmach*, a politico-military title which literally means 'commander of the right',[11] for example, was bestowed on Kalla by Emperor Haile Selassie himself. The regional *poqallas* who became *balabbats* also received cloaks and fine clothes similar to those worn by other dignitaries of Empire. In the 1960s, Haile Selassie visited Konso by helicopter and put up his tents in the forest next to Kalla's homestead. This forest is said to be sacred: no one can cut any tree without Kalla's permission, and it is where his forefathers are buried. Haile Selassie stood on a high cliff at the edge of this forest and made a speech to many Konso people who had gathered below to hear him, and clothes were distributed to the people. The way in which the Emperor stayed beside Kalla, instead of in the administrative headquarters of Bekawle, shows the way in which Kalla was recognized as important, and the way he had cultivated relations with the northerners.

Bamalle, whose homestead was closer to the administrative headquarters of the northern administrators, also cultivated good relations with them, and obtained new powers through his position as *balabbat*. He seems to have taken an active role in collecting taxes from people, and was particularly fierce in collecting a tax from those who used the extensive lands that he held below the town of Dokato. His claim to much of this land was probably ancient, but he may have been able to add to it through his new role. When I first interviewed Kaitari Bamalle, he told me that during the time of the northerners he 'exploited the people terribly'. He refused to elaborate on this later, except to complain that:

> When the northerners came they took over our authority. We had to pay taxes to them. Before people paid taxes to us. If we went to someone's house and there was a fat goat there then they would give it to us. If there was honey they would give it.[12]

He saw the northerners as usurping many of his customary rights and that it was therefore only fair that he use his position as *balabbat* to regain as much advantage as possible. It is said that if someone collected honey from a tree on Bamalle's land during the time when he was a *balabbat*, and failed to pay tax on it, then he would have been killed.

The families of Bamalle and Kalla also both converted to Orthodox Christianity, the religion of the conquerors and of the Empire. Kalla gave his children Orthodox Christian names, and sent them to school to learn to read and to speak Amharic. It is possible that it was at this time that the mythical place of origin of the Kalla family, Kontari, was equated with Gondar, a great historic centre of northern Amharic culture and religion. The impact of the incorporation of Konso into the Empire on the position of regional *poqallas* is thus similar to the history described by Donham and

Markakis. It increased the social difference between the regional *poqalla*s and other people. The lands of the regional *poqalla*s were already significant and dispersed throughout Konso, but it is likely that these *poqalla*s were able either to secure or to add to their landholdings, at this time.

At a village level, a different picture emerges. The story of what happened in Buso cannot be considered to represent what happened in all villages, but it is significant because of the way in which it deviates from the pattern suggested by Donham and others working in this area in this period. In Buso, the *poqalla* Aija was made *balabbat*. When I asked what this meant I was told, 'all the land was said to belong to him, and the people had to pay rent to him, in taxes.'[13] But all those I talked to were at pains to state that this was an arrangement in name, for administrative reasons: although all the land *was said* to belong to Aija, who was *balabbat* and collected taxes as rent, 'the people who owned the land before still had control over it and managed it in the same way, and land was freely bought and sold.'[14] Those I interviewed in the field, *poqalla*s and others alike, also comprehensively denied that the local *poqalla*s gained any land from their work as *balabbat*s. Instead they stressed that their control of large amounts of land dated from the time of their settlement in Konso. For example, a member of the *balabbat* Aija's family explained:

> the position of *balabbat* had no salary, and no gifts were given to Aija for this work, and they did not increase their land at this time, they had a lot of land before. Being a *balabbat* did not change this. Once, someone said that they were sometimes given special clothes to wear, but there were no other material benefits. The only thing was that if he went to a house where there was *chagga*, then he could drink some freely, in the same way as the *apa para* or *apa timba*.[15]

Of course, it is to be expected that the *poqalla*s' families would say this. Such claims naturalize and strengthen their control over large amounts of land. However, this account was supported by other non-*poqalla* members of the community and no one contradicted it while I was in the field. The view is exemplified by the description that one elder gave of *poqalla*s:

> They were the first to come to this area, so they became a big family, and they became *balabbats*. The *balabbat*-hood came together with the northerners (*qaweta*), but long ago they were kings (*negussada*).[16] They were kings of this land. Even when war comes to this area, they go and then it stops.[17]

One point particularly emphasized by all interviewees on the subject was that the village-level *balabbats* in Konso could be dismissed on complaint by the people. This prevented them from abusing their power. This is also confirmed by Hallpike (1972) who worked in the area at the end of the period of northern rule, and was able to witness the situation at first hand. He writes that 'all said that if they did not like their *balabbat* they could always sack him and choose someone else' (ibid.: 71). He also described the limits of *balabbat* power:

> these functions confer no power on the *balabbat* within Konso society except in so far as his responsibilities to the administration enable him to bring the police to arrest criminals and those who refuse to pay their taxes or attend court (ibid.: 66).

At a village level it is also noteworthy that there is little difference in the landholdings of *poqallas* who worked as *balabbats* and the *poqallas* who did not. Aija has large amounts of land, but his landholdings are the same order as the landholdings of *poqallas* who were not *balabbats*. The large landholdings are more likely one of the reasons why he was considered senior and chosen to be the *balabbat* than a result of his *balabbat* position. As Hallpike wrote again, '*[b]alabbat* families were always the most eminent priestly families in the towns and clearly the Amhara administration has chosen them as the ones most likely to command respect' (ibid.: 71). As *balabbats*, they would have been able to maintain their control over their land, but their control over large amounts pre-dates colonization.

If one of the defining features of being class-like is having control over large amounts of land, then, at a village level, the relations between *poqallas* and others were not transformed by the incorporation into the Empire into becoming more class-like. [18] But it is necessary to look more carefully at the *nature* of the control over the land in Konso to confirm this argument, as Donham argues that, in Maale, it was only through the incorporation process that the control over land became more *exclusionary*, and therefore the basis for more class-like relations.

The Konso claim that, long before the northern rule, land was individually held, and was alienable: the owner could give it away, leave it fallow, cultivate it or sell it, as he pleased (and as women do not have the same land rights, it is possible to say 'he' in this case). Exclusive rights to land tend to develop with the permanent cultivation of fields and with the investment of labour in the construction of landesque capital (Netting, 1993; Blaikie and Brookfield, 1987). In Konso, therefore, there was no similar shift in the nature of rights to land because, as discussed, exclusive rights to land had already developed.

The *wakas* support these claims. They show that land has been bought or exchanged for a long time, and they show that there was a strong identification of the individual with the pursuit of the acquisition of landholdings. But the *wakas* also emphasize a unity between the *poqalla* and his descendants, and between an individual's interests, the group's interests, and those of the *poqalla*. For example, those persons said to have been killed by the *poqalla* had normally been killed by a member of the lineage. Similarly, the *poqalla* appropriates the prestige associated with the kills that his lineage makes of wild animals. Most significantly, the different kinds of stones that are used for the different kinds of land acquisitions (kin and non-kin) suggest that the land acquired from outside the lineage was more prestigious, as it was viewed as adding to the wealth of the lineage as a group. The land may ultimately have been recycled as it was given to the descendants of the *poqalla* as they were born. This distinction between the lineage and the non-lineage suggests that something approximating the corporate group existed, and that the lineage exercises some rights to the land of the *poqalla*. This puts in doubt the theory that exclusive rights to land developed before northern rule and demonstrates that there may also have been a shift to more class-like relations with northern rule, as in Maale.

But although the lineage may have certain negotiable claims on the resources of the *poqalla*, this does not necessarily equate with the existence of a corporate group in terms of shared interests or shared rights to property. Most of those who 'breastfeed' have little choice: it may be the only way for a younger brother who inherits little to obtain land, or goods for marriage, or food. If he does not want to 'breastfeed', because he does not want to comply with the obligations that accompany the process, the alternative is to leave Konso for the surrounding frontier. A quote from Kahano Kalayta and Kahano Kolobo demonstrates that lineage members are aware of the consequences of 'breastfeeding', and they do not think of it as a shared project:

> Some poor families make relations with big families [*poqallas*], and through this method they try to improve their lives, through coming close to them, the big families. Through being close to the big families they extract something useful for themselves. Others say to themselves, I was poor before and I am poor now, and I must struggle for myself to improve it. If I become weak, I become weak, I still will not rely on others. The one who makes himself close to the big families, through them his children grow up, through this they can do their work, as before they were too poor even to cultivate their fields. Small families, through making relations with big families, may receive a small and fertile terrace to use. Also, the small families can take blessings from the big families. Small families who refuse to make relations with big families, then they have to leave for Gummayde, or Gidole or Fuchucha, and they are chased to there. What can they do? They can't do anything. They don't have anything. It is like that.[19]

Secondly, the lineage members, like other people, may borrow land, but they have little or no other rights to the land of their *poqalla*. In the research, only two cases were encountered in which the lineage members were trying to influence what the *poqalla* did with his land. In the first a *poqalla* had been in a dispute with his lineage as the *poqalla* had reputedly failed to give land to his brother that had been bequeathed to the brother by their father. The lineage members called elders in the village to advise the *poqalla*, and they refused to work on his land, to consult him, or help him in any other way until he had given his brother his rightful land. This case demonstrates that the lineage had some influence over the use of the land of the *poqalla*, but it is not evidence for the existence of right to decide what happens to the land itself: the lineage members did not make claims to any new land; they were only condemning the *poqalla* for failing to honour his father's promises. The second case refers to a *poqalla* who had many problems. He drank a great deal and spent most of his time searching for casual labour (in Konso and further afield). He sold land cheaply, in order to obtain money quickly. His wife, with the support of the lineage members, was trying to stop him from selling his land, even though they acknowledged that according to custom it was his right to do so. Their problem with his activities was that he was selling land far below its value, and for short-term ends. They described him as an *'almala'*: a word in Konso that means a good-for-nothing and translated by Black and Shako (1973) as someone with no social responsibility. Again, therefore, this is not so much a case of the lineage having rights to the land, but a case whereby the lineage members were trying to negotiate with a *poqalla* who they felt was behaving irresponsibly.

These examples are of limited use because they are from recent times, but I believe that similar conflicts and contestations within the lineage may have existed before. If this the case, it is possible that relations between a *poqalla* and his lineage, although sacred and mediated by ritual, always showed elements that may be described as class-like. The relations between a *poqalla* and those who are not his lineage can be seen as even more class-like. Unlike the Maale case, I argue that in Konso, more class-like relations already existed. This is not necessarily surprising, as other writers, particularly those described as the French Marxists, writing in the 1970s-1980s, and mentioned later, concluded that classes were often present in societies organized by kin relations and in which the corporate lineage was present. These writers also stressed the importance of controlling prestige goods (or symbolic capital) as much as the material means of production (or economic capital) for maintaining the position of the dominant class:

> there therefore seems to be a general consensus of presence of classes... in lineage societies 'when the extortion of surplus labour from the direct producers permits the dominant minority, either directly or through exchange to acquire prestige goods, the control of which defines its power (Kahn, 1981: 78 [quoting Terray]).

To some extent it is a matter of perspective. Drawing a line between a relationship of inequality and a relationship that serves certain useful functions depends on one's experience, which may change over time. It depends on the ability of an individual to make use of opportunities provided by the structures in which he or she is positioned. What is clear is that some are more advantageously positioned than others. The two ideal types discussed earlier in this chapter may not be exclusive either. In Konso, the 'sacred hierarchy' mediated by ritual and shared forms of identification coexists with more class-like relations based on unequal access to and control over resources.

Like capital, class is a problematic concept to use because it has developed in the Western industrial and capitalist experience and brings with it associations and assumptions that cannot necessarily be applied to another context. It also implies the existence of well defined homogeneous groups, and ignores the existence of multiple axes and forms of inequality that may cut across each other (McDowell and Sharp, 1997). It may bring with it assumptions about the existence of 'class consciousness' or the ability for a class to mobilize and resist the relations of inequality. But these notions are not necessarily appropriate to the Konso case. A simple definition of class as 'positionality in relation to capital circulation and accumulation' (Harvey, 1998) is more useful, especially when it is stressed that classes do not exist as independent entities or categories, but as forms of social practice in which inequality is produced and institutionalized. This idea of class is similar to what Godelier describes as Marx's 'generic' use of the term class, that is applicable to non-capitalist as well as capitalist societies (Godelier, 1986).[20] Such an idea of class is useful, because it limits itself to identifying, describing and understanding differential factors that have powerful and real effects on a person's life experience (Ortner, 1998; Harvey, 1998). The individual's positionality in relation to the different forms of capital shapes the constraints under which he or she lives, and also enables him or her to take advantage of the opportunities that are presented.[21]

The different components that shape the class-like relations in Konso include, therefore, the differential control over land to which exclusive rights are held, and the differential control over prestige goods and symbolic capital. As this symbolic capital is seen as bringing fertility, it could also be understood as a control over the means of reproduction, which, according to Meillassoux (1981), is as important in maintaining inequality as controlling the means of production. Donham (1986b), Meillassoux (1981), and Bourdieu (1977; 1984) also agree that different lifestyles, dispositions and social practices are as important in creating difference and maintaining it between different groups as the control over capital. In Konso, the lifestyles of *poqalla*s are different from others, often markedly so, because of their observance of behavioural taboos. Perhaps this is nowhere more evident than in the case of the regional *poqalla*s who are forbidden from cultivating the fields themselves (a behavioural taboo which can also be seen as a symbolic manifestation of the *poqalla*'s access to the labour power of others). There is also evidence of preferential marriage between *poqalla*s, which, according to Meillassoux (1981), is one of the most important expressions of class divisions, and ways of re-inforcing them.[22]

The different position and lifestyles of *poqalla*s and others is also played out through the organization of space, and different degrees of control over particular spaces. Regional *poqalla*s live apart from other Konso. They also 'own' symbolically important places such as *moras* (public meeting places) and sacred forests. Successful local *poqalla*s also take on some trappings of a spatially defined position: they 'own' *moras*, they have large houses, and the most important live on the outskirts of villages, claiming to be 'first' in the village, thus gravitating towards the sort of spatial position of a regional *poqalla*. This increasing territorialization of authority in proportion to the degree of stratification of relations fits in with Sack's schema of the increasing spatiality of power with its development: control over space is equivalent to social power (Sack, 1986). The amount of land controlled by the *poqalla* can be seen to be proportional to his influence and power. In some cases (particularly in the case of regional *poqallas*) these social differences, which were already present, have been made more extreme through the process of northern rule.

When comparing with the Maale case, the important point is that, in Konso, these class-like relations were not produced out of the incorporation into the northern Empire, but have a longer history. Social differences and control over land may have been strengthened through incorporation, and made the class-like relations more extreme. The idea that class relations can be more or less extreme informs the understanding of the power relations between men in Konso more generally: there is a spectrum of positions of power, and the power relations between people on this spectrum may be more or less class-like. At one end of this spectrum are the regional *poqallas*, while at the other end are the ordinary descendants – the people who hold little land and no sacred office. The distance between people on this spectrum is associated with the level of the development of the class relations. Those in different groups on the spectrum cooperate, with various degrees of inequality, to sustain livelihoods, but inequality never disappears. The idea of a spectrum is useful for envisaging these power relations because the spectrum itself has not been radically transformed, but may have been stretched, by the process of northern rule. The wider the distance between different people on the spectrum, the more class-like the relations between them.

The nature of relations in Konso was, therefore, not radically transformed by incorporation into the northern Empire. Another reason for the difference from Maale could be the fact that the northerners found Konso such a stony and hilly place that they were not greatly interested in intervening in the landholdings. At the local level, the Konso were able to maintain many of their customary arrangements.[23]

Metaphors of power

Some of the metaphors that the Konso use to think about these power relations are useful to explore more emic ideas about how these power relations are structured and experienced. Metaphors are more than just descriptors; they shape everyday thought and action (Lakoff and Johnson, 1980; Tilley, 1999). Metaphors are also loaded with values and feelings (Fernandez, 1991). If metaphors are deconstructed, the meaning and structures of different understandings and visions of the world can be examined. Barnes and Duncan emphasize this by asserting '[b]y our metaphors you shall know us' (1992, 12). The Konso metaphors that are used to talk about power relations show the way in which the unequal relations are naturalized and institutionalized. However, the metaphors used for describing unequal relations also reveal that the relations of inequality between *poqallas* and others serve as a template for unequal power relations that exist at other scales of society (and vice-versa). Exploring these connections can therefore provide a comment on the other axes of inequality that exist in Konso, as well as those between *poqallas* and others.

The *poqalla* refers to his lineage as his 'children of the clan' (*xella kaffa*), his urine (*sindha*), or as his little brothers (*kussitas*). The first metaphor that likens the lineage to children repeats the nurturing theme seen in the metaphor 'breastfeeding'. This theme is also present in the rituals that accompany the inauguration of another form of authority in the village: the *apa timba*. When the new *apa timba* receives the drum and the office, he is told to '*ino umbani, ino toyiti!*' which means, 'may you hold us safely, may you watch over us'. The verb '*umbani*' refers to the holding, rocking and comforting of a baby in one's arms. When an authority is told '*ino umbani*', he is being requested to act like a mother, father, or elder child, who protects and soothes a younger one. When a *poqalla* refers to his lineage as his 'children', it is this kind of power that is being referred to. Bourdieu (1977) and others (e.g., Gledhill, 1994) would see these metaphors as disguising the extractive nature of the power relations that have been discussed in previous sections. The fact that metaphors like 'child' or 'breastfeeding' may be 'euphemisms' (Bourdieu, 1977) does not diminish their power. These metaphors have the very real effect of naturalizing the inequalities and presenting the power relations as a benign and protective force that exists for the good of all. There is no doubt that their power is partly derived from the way in which the elites may protect and provide, as well as extracting, labour.

'*Sindha*', the second metaphor that refers to the lineage as the *poqalla*'s 'urine' is more obscure, and those interviewed were less willing or able to expound its meaning. The most obvious interpretation would be to equate urine with semen, and to see the metaphor of urine as referring to the way in which the lineage as descendants of the *poqalla*, have literally flowed from the *poqalla* himself. The *poqalla* is the embodiment of fertility; fertility flows in semen; urine like semen flows from the penis (and there is therefore an association made between these fluids). This interpretation is supported by Sperber's analysis of Dorze, in which Dorze ideas about seminal flow and retention are, he argues, often expressed at a '"pre-genital" level' (Sperber, 1975: 40). Seminal flow and retention are paired symbolically and metaphorically with the expulsion or retention of other bodily fluids, particularly faeces and urine. In Dorze, dignitaries urinate at key moments in ritual, and in Konso dignitaries are often specifically forbidden from urinating until a ritual has been completed.[24] Referring to the lineage as the urine of the *poqalla* therefore stresses the way in which they have been born from him (as he represents the founding ancestor). It also emphasizes the way in which he is the embodiment of fertility of the lineage, a fertility that can be interpreted in a narrow sense, or in a broader one, to refer to rain and wellbeing as well as to biological reproduction.

The third metaphor for discussing the lineage is that of 'little brothers'. One man, for example, explained of his relationship to his *poqalla*: 'They say I am like his little brother, so I will take milk from his breast, that's why his fields I dig as if they were my own.'[25] When the lineage members are referred to as 'little brothers', *poqalla*-lineage relations are equated with elder–younger brother relations at the household level. The relations between brothers mirror at a larger scale those between a *poqalla* and his lineage in various ways. The elder brother has seniority over the younger brother because he is born 'first', in the same way that the *poqalla*'s position and authority derive partly from his claims to firstness. The elder brother also inherits much more than a younger brother: the elder brother is a receptacle for the name and land of the family in the same way as the *poqalla* is for the lineage. This inequality is institutionalized and unalterable, as it is fixed by birth and is emphasized in the naming of second brothers as 'Kussi'. A younger brother who inherits little may need to call upon his elder brother for assistance: for extra food, perhaps for the loan of a field, for a cow skin to sleep on, for wood from his brother's fields to build a house, and for labour for all of this. In return for this assistance, the younger brother is able to help the elder brother in his fields. One man explained the relations of elder brothers to younger brothers in these terms: 'when he [the elder brother] needs something he sends them [younger brothers]. This is what respect means.'[26] This demonstrates that with the seniority of the elder brother come some rights to the labour of his junior brothers.

It might be expected that the division between the lot of the elder and younger brothers might cause a lot of conflict, but in practice it seems to be remarkably easily accepted as the way things are – perhaps like all strong social and cultural organizations. Occasionally, a younger brother feels bitter towards his father, and I was told that it is common for a younger brother to ask his father 'Why are you making me so poor, and him so rich?'[27] A younger brother might sow his crops over the boundary of the field that he shares with his elder brother, forcefully encroaching on his inheritance. This behaviour usually results in violence between the two brothers. But such resistance, or 'weapons of the weak' (Scott, 1985), is rather limited.

During the time of field work, accusations of witchcraft were very rare, but Hallpike noted that in the past they were not uncommon, and that brothers did not like to live next to each other in the village because of fears of witchcraft (Hallpike, 1972: 127). This suggests that witchcraft was one way in which anger against the inequality could be exercised.[28] The inequality between brothers is such a basic and fundamental dimension of Konso life, that when other cases of inequality in Konso life are encountered, the relationship between brothers is evoked to explain the inequality. For example, if one village is more important and powerful because it is bigger in population and wealth, then the Konso will say, 'you see, the big village is more important because it is like the elder brother to the little village'. The relationship between artisans and farmers is also explained through the metaphor of brothers: artisans are considered junior to farmers, and when asked why, it is often said that 'you see the farmers came first, they are like the elder brother to the younger brother'. The relationship of brothers is evoked at once to explain and to naturalize the inequality.

At slack periods during the farming calendar, the Konso occasionally play a game at night in the *mora* called *kureila.* A puck is thrown in the air and two teams have to compete to catch it. Although the village is divided into different wards, the teams that play against each other are not of different areas, or different generation grades, or any other groups, but of elder against younger brothers. The atmosphere surrounding the game is one of carnival: men and women, young and old, all come to watch, although in usual circumstances, a woman should not go near the *mora* at night. It is possible to see that this is a ritualized scenario, and a context in which it is possible for the younger brothers to challenge the elder brothers openly. Most of all it shows the importance of these two categories in Konso thought and organization.

However, although the elder brother inherits the majority of the property of the family, there are some disadvantages to being an elder brother, and advantages to being a younger. For example, the elder brothers have to live with and look after their parents and there is often conflict between the incoming wife of the elder brother and her mother-in-law. The family also often puts pressure on the elder brother to marry and have children at a very young age, to continue the family line, whereas the younger brother may do as he wishes. Younger brothers may go to school, but it is thought that elder brothers should stay at home and look after the fields. So the life of an elder brother can be full of the expectations of others, and some restrictions, whereas a younger brother can do more as he pleases.

The idea of the relations between brothers as a model for the concept of inequality is a strong demonstration of the way in which the inequality is thought about and experienced in Konso. This can refer to the inequality between brothers, or between *poqalla*s and others or between farmers and craftspeople. The authority of the *poqalla* and the relations between a *poqalla* and his lineage are therefore modelled on the relations between brothers, which in turn are based in claims to being first. This modelling of inequality on brothers naturalizes it and legitimizes the power and position of the *poqalla*. The use of kin relations as a metaphor for power relations makes the inequality particularly difficult to challenge. The inequality is recognized as inevitable, but there are also strong feelings of solidarity, in that they were born of the same parents, share the same blood, and that benefits are to be gained from cooperating with an elder brother. At the same time, although the *poqallas*' authority is modelled on brothers, the *poqallas* also provide a sort of ideal that other men should strive for, so the modelling works both ways. Relations between brothers may also be modelled on relations between *poqallas* and their lineage.

The relations of inequality in Konso are thus thought to be natural. They are naturalized through claims to firstness, and through kinship and other metaphors. I came across other bodily metaphors which are also commonly appealed to in order to explain and justify these inequalities, and the place of different people in these inequalities. For example, I once asked a wealthy and influential craftsman if his wealth and importance made him equal or senior to a farmer. He dismissed this idea saying '*xolma mata ijalinjan*' – 'the neck cannot come over the head', showing how these relations of inequality are thought of as fixed, and that the positions of being more junior and more senior cannot be – or should not be – challenged.

Relations between men and women

I did not come across metaphors that are used specifically to refer to the relationship between men and women. Women, who have no land of their own, are said to be 'below' their husbands in the homestead. Many stressed that women, who marry into their husband's home, will always to some extent have outsider status. The only thing that can transform a woman's position from this marginal status is if she gives birth to a son, as women are considered to have the right to be respected in their son's home.[29] If a woman gives birth to a son, she is often referred to from that time after as the 'mother of so-and so', rather than by her given name. If a woman does not give birth to a son, then her position often remains marginal.

The inequality that exists between men and women is also used as a metaphor for other axes of inequality. The inequality between farmers and craftspeople, for example, which is often explained in terms of brothers, is sometimes referred to in terms of the different genders: the craftspeople are said to be the 'sisters' (*tubarteta*) of the farmers.

The difficulties that women experience as a result of their position should not be underestimated. Their work is hard. During a lifetime they have many pregnancies and they often experience difficulties in childbirth; child and maternal mortality rates are high. The relations between men and women are – like other power relations – influenced by the structures

Photo 4.3 *Buso women in focus group*

of power that have been discussed above, but they are also mediated by the nature of the individual relations between the people involved. In the case of relations between men and women, where individual feelings and dispositions have a particularly strong influence, it is difficult to generalize about the nature of the power relations. Women's own words reveal more about their position and experience. One interview with a woman who had been recently elected on to the village committee captures many aspects of her life (questions are included in italics):

> *How is the life of women in Konso?*
> The life of women in Konso is very difficult. We collect fodder for the animals, go for water, cook and, like men, we collect stones and bring them home to build our homes, and grass to thatch them.
> *Who makes the decisions in the home?*
> Father/husband. He is the head of the home.
> *Who makes decisions about the grain?*
> I will decide on its usage. If I am a good wife, I will use it wisely. If I am not, it will be used in a little time. After I have cooked, I give the food to father first. Then I give it to the eldest son, then the second son, then the other children. I just wait and then I take whatever is left.
> *What if the food is little?*
> I still eat what is left over.
> *How is that for women?*
> Women are very, very hungry.
> *Are there other ways of influencing what happens in the home?*
> No, if I do anything secretly without his permission he will hit me or chase

me from the home. When I make *chagga*, I must give it to him first.
So how is the life of women compared to the life of men?
The life of women is very hard, but the life of men is also hard. But we are carrying firewood; we are going to the farm to collect food for the animals; we are cooking. If we don't do some of this then we will be beaten by our husbands. So our lives are somehow serious.[30]

Other interviews followed a similar pattern. Women usually listed the number of their jobs and their responsibilities. Many complained that if they had to go far to collect firewood for cooking and brewing, they could not care for their children properly. Two women I interviewed, for example, had spent seven hours on the road on the previous day collecting grass to use as firewood for brewing (two hours' to the fields; two hours' cutting the grass; three hours' return). Both the women were breast-feeding, and the babies had to be cared for by young children while they were away.

Another group of women explained the situation of women:

> When the husband decides something we accept it. And since we are women we can't override his decisions. But we give advice to the children and tell them what to do. We are responsible for the children in the home.[31]

And they commented on how 'if they make a mistake,' then their husbands may beat them: 'If a wife makes a mistake then he'll become very angry and the wife will be afraid. When this happens if you are not quiet then he'll hit you'.[32] And when I asked what counts as a mistake, I was told:

> For example, if I go to market, I might meet up with a friend, and then we might stop and drink a gourd of *chagga*. Then we'll stop a little time and I may become late. If he comes back from market before me then he'll be angry. If he hits me, I can go to my family and tell my brothers and my mother. Then they'll ask the reasons for his beating me. If there is a reason then they'll advise me to return to the home.[33]

I witnessed a number of appeal cases (in various forums) on behalf of women who had been beaten. These cases were disturbing, but what was most disturbing was the way in which beatings were often considered to be legitimate, by both men and women.

As with the other power relations in Konso, the power of the senior party in the relationship is significant, but not absolute or unlimited. For example, women have some control over the food in the granary, and this gives them some authority within the household. They are considered the main carers and have the main responsibility for the children. As with other power relations in Konso, it is important also to encourage the co-operation of the less powerful party, and this sometimes prevents power from being wielded too harshly. Women's labour is particularly important and because of this it is important also to give women some respect. In order to call a work party, it is necessary for the woman of the household to brew beer for the workers. The importance of obtaining access to this labour gives the women some bargaining power. The power relations are unequal and unjust, but also cooperative and to some extent functional. This dual quality can be extended to understandings of all the relations of power that have been described in Konso: between *poqallas* and others, between brothers, and between men and women.

Conclusion

This chapter has examined the nature of the position and the power of the *poqalla* in more detail. It has shown that the *poqallas* exist in a privileged but precarious position in relation to the control over the major forms of accumulatable capital, land and symbolic capital. They use their privileged access to these forms of capital to obtain access to another form of capital that is most crucial to the intensive agriculture – and which is also the most difficult to accumulate – labour. The *poqallas* act to produce and reproduce the intensive agriculture of Konso, and in this way they are 'adaptive' in that they help to protect livelihoods, to minimize risks, and to conserve soil and water. However, they are also a form of political centralization and are forms of an institution engaged in competition for power. The adaptation to a given environment, and the transformation of it is not separate from the struggle for power of elites. Similarly, in exploring the unequal power relations between *poqallas* and others, a degree of protection and provision (for others) goes hand in hand with a more privileged position (on the part of *poqallas*) and their struggle to acquire the labour of others. This does not mean that unequal relations and a degree of centralization of power are always the case with intensification, and with the adaptation to particular environments that it represents. Nor does it necessarily justify the existence of unequal power relations.

This chapter has also demonstrated that the *poqallas* exist as a kind of template or instructive example for other unequal relations that exist in Konso. The other relations of inequality that exist, between brothers particularly, but also to some degree between men and women, mirror at a smaller scale, and also to a lesser degree, the relations between the *poqallas* and others. In Konso, as in Maale, there are 'inequalities pyramided upon inequalities'; there are power relations that are nested within other power relations like Russian dolls. These power relations to some degree appear to be connected to the agriculture of Konso, in that a key dynamic of them is the need to secure access to the labour of others, and to use other forms of capital in order to do that. Control by men over women's bodies and women's lives, in particular their wives, is another aspect of this dynamic. These forms of control and relations of inequality are institutionalized in metaphors and other structures such as the *wakas*.

The reason for exploring these relations of power is in part to make the point that these *poqallas* – institutions that play a role in the maintenance of the landscape and livelihoods – are political institutions. They are involved in struggles to maintain control over people and land and other resources. When development organizations are looking to find indigenous institutions that play a role in NRM, it is easy for them to concentrate on identifying the roles that institutions play and to forget to examine the relations of inequality of which they are part. This is not to say that this means that partnerships should not be made with indigenous institutions which are likely to be elites of various kinds, but if a development project does decide to form an alliance with these kinds of institutions it is important to be aware of the nature of the institutions involved, and to know the history of their position and power. These institutions do not exist as unproblematic representatives of their communities, but are institutions of power which have emerged out of particular histories. By working with these institutions, precautions must be taken – if possible – to make sure that the relations of inequality which they represent are not further entrenched. The information presented in this chapter demonstrates that partnership with indigenous institutions like *poqallas* may strengthen the position of the indigenous elites, at the cost of the poorest and those at the bottom of the 'pyramid' of unequal power relations. Chapter Three showed that *poqallas* frequently lend land to poor women. This chapter has shown that the structural position of women overall could also be worsened by development initiatives that strengthen indigenous institutions, as these approaches also strengthen the structural relations of inequality.

Notes

1 'The acquisition of a clientele, even an inherited one, implies considerable *labour* devoted to making and maintaining relations, and also substantial material and symbolic investments, in the form of political aid against theft, offence and insult, or economic aid, which can be very costly, especially in times of scarcity' (Bourdieu, 1977: 180).

2 Those who described this process of taxation were vague about the period during which is was applied. They implied that it was an arrangement that predated the colonial period, but it is possible that it was a privilege that these *poqallas* claimed after they were given certain administrative duties by the northern empire (see later).

3 The idea of an 'accountancy' is borrowed from the work of Bourdieu, who claims that 'an accountancy of symbolic exchanges' is necessary in order to understand the nature and workings of power (Bourdieu, 1977: 178). Symbolic exchanges, as I will show, are important here, but the mundane and ordinary material exchanges must be analysed too, as Bourdieu's work would confirm (*ibid*).

4 There are inherent limits to these processes, and the dynamics of the size of a clientele may also be a factor: if the clientele is too large or too small it may be difficult for the *poqalla* to effect the virtuous circle.

5 For example the *tuuta* in Fasha where all pests are cursed.

6 In Fasha area, a similar ceremony is called '*tuuta*'.
7 Watson, 1998.
8 All of the italicized terms referring to the policies institutionalised by the northerners are derived from Amharic (unless otherwise specified), though many have also entered the local language.
9 Interview with Kembo, 2002.
10 Interview with Rakasa of Dokato, 1996.
11 Bahru Zewde, 1991: 233.
12 Bamalle, 29 January 1996.
13 Kahano Kalayta and Kahano Kolobo, 1 December 1995.
14 Kahano Kalayta and Kahano Kolobo, as above.
15 K.A., wife of *poqalla*, interviewed 1995
16 The word he used for king is derived from the Amharic. There is no Konso equivalent.
17 Kairiti Tenayu, elder from Gaho village, 1996.
18 Money does not seem to have been an important form of capital at this stage. People did have some access to money; those, for example, who fought for the Italians discuss the amount they were paid. Money becomes an important accumulatable form of capital much later and has a strong transformational power in the 1990s (see Chapter Seven).
19 Kahano Kalayta and Kahano Kolobo, 1 June 1996.
20 Pages 245–52 of Godelier, 1986, review the different meanings that Marx attached to the term class.
21 This is consistent with an approach to power that sees it as both constraining and enabling (for example, Lukes, 1986).
22 In Konso, a regional *poqalla* must marry the daughter of another *poqalla* (usually a local *poqalla*). The girl is chosen while she and the *poqalla*-to-be are still children. Among local *poqallas*, the information collected showed a tendency for *poqallas* to marry the daughters of other *poqallas*, forming alliances between *poqalla* families.
23 Amborn writes of the Burji-Konso cluster that 'at least some part of the land was not important to the northerners. The steep terraced slopes of Konso were not suitable for the cultivation of *t'eff*, and neither could the plough be used there' (Amborn, 1988: 755). This general picture is confirmed by Markakis who presents figures to show that relatively little land in Gamo Gofa (where Konso was located) was divided up under the *sisso* principle, and formally divided and distributed to people. His figures for the percentage of *sisso* land in the different regions in 1968 are, in decreasing order: Illubabor, 15.6%; Wallega, 13.70%; Arusi, 11.7%; Shoa, 9.8%; Kaffa, 2.29%; Sidamo 1.32%; Gamo Gofa, 0.06%. Although these figures were recorded in 1968, they can be assumed to reflect the levels of intervention in landholdings over a longer period.
24 In the *xoxayta* ceremony, the *shorogota* must stay awake all night cursing pests and 'bad things'. During this time they cannot urinate (and this point was stressed as important by all the *shorogota* interviewed). When there is no rain, the people of Turo region go to the regional *poqalla*, Kalla, and ask him to pray for rain. Kalla carries out a ceremony and blesses the men who have come. When they return to Turo they must not urinate until they have crossed the river that marks the boundary between Turo region and Karate region (where Kalla lives). This also demonstrates that urine does not only represent semen, it represents, according to Sperber (1975), conspicuous expenditure in a broader sense that takes place in rites of passage. In the rain ceremony it symbolizes rain – another fluid necessary for fertility – and when a man from Turo failed to heed Kalla's advice and urinated before crossing the boundary river, floods came and washed many people away.
25 Kairiti Tenayu, elder from Gaho village, 1996.
26 Apo Passana, elder from Buso village, 1996.
27 Questionnaire number 23.
28 Donham's work on Maale also shows how brothers who had unequal inheritance often consulted diviners when relations between brothers became sour (Donham, 1990). He also shows how younger brothers could not effectively resist the inequality, as the Maale thought it was 'natural' for the elder brother to control the production (and

reproduction) of the younger brothers, both ritually and in practice. If a younger brother went against this, then he was transgressing the established order and likely to be punished with poor crops and barrenness. Donham describes how a younger brother who does go against his elder brother, will sooner or later experience some misfortune, and the community will conclude that the younger brother brought this upon himself because of his resistance:

> If a younger brother who infringed the rights of his eldest continues to live in the same area... every misfortune in the brother's life (and given the material organization of Maale society, there will perforce be misfortunes) will be interpreted by those around him as a consequence of his resisting proper control by his eldest brother. In the end, as his child lies dying or as locusts destroy his field, an estranged younger brother will almost have to present a compensation payment to his eldest brother in order, as the Maale believe, to prevent further tragedy. And the payment may be greater than the original bone of contention between the two. (ibid.: 99)

29 The rights that are gained by giving birth to a son are made most clear when the woman dies. A woman who has given birth to a son can always return to that homestead and should be buried by that family and mourned by them, even if she left a long time ago and has married subsequently elsewhere. Such rights are not extended to women who give birth only to daughters. If a woman leaves her husband, then the children must remain with their father, and this is rarely challenged.

30 Member of village committee, Buso, 2002.

31 Discussion with group of women, Buso, 2002.

32 Discussion with group of women, Buso, 2002.

33 Buso woman, 2002.

Five

Modernity & Christianity

Chapters Two, Three and Four examined the way in which the production and reproduction of the landscape in Konso is part of a social, cultural, ritual and political process of which the *poqallas* are at the heart. The power of the *poqallas* rests in their control over and embodiment of forms of social, symbolic and economic capital. The study of the *poqalla* shows that the institutions for managing land and labour are embedded in other aspects of society to the point that the economic, the ritual, the judicial, and the political cannot be separated. This raises questions about the usefulness of the term 'embeddedness', often used to refer to non-economic aspects of production in economics. The term 'embedded' implies that the economic (or productive) practices are grounded within existing social networks and cultural beliefs and practices, from which, therefore, it could be separated. This research shows that the economic, ritual, social and political are mutually constitutive. Everyday life in Konso (and elsewhere) is made up of interactions and exchanges between people which are at once social, cultural, ritual, economic and political; it is difficult to prioritize any one dimension of an exchange over another. To separate them would be to impose dichotomies and divisions between different sectors that do not exist in practice.

The last chapter compared data from oral histories, from burial statues, and from the political events from the 1890s, to argue that the role of the *poqalla* in the production of the landscape is ancient. The *poqallas*' position may have changed over the twentieth century through their role as intermediaries with the state, but their role in production and their control over land is not new. The unequal relations between *poqallas* and others are not entirely new either. This stress on continuity in the role of indigenous institutions over time is important, especially as much work on the African continent stresses the way in which 'traditions' have been invented, often through the actions of colonial administrators who attempted to harness indigenous institutions for state ends (Iliffe, 1979;

Mamdani, 1996; Ranger, 1983). The case in Konso emphasizes that it is important to take into account the continuities of indigenous practices and power relations, as well as the inventions.

But the previous chapters tell only part of the story. It would be possible to leave it at that, with the light of analysis, like Eliot's candle in the quotation at the beginning of this book, illuminating only one aspect of what has taken place in Konso. Over the last century, there have also been other radical changes and transformations that have challenged the power and legitimacy of the *poqalla*, and that have also provided new and alternative forms of organization. Konso is not an 'island', literally or metaphorically, to draw on the words of Hardy also quoted at the beginning of the book; it is connected to a 'great web' of flows of change, economic, political and religious. These have brought radical changes. The aim of the following chapters is to explore and to understand the nature and impact of these changes and transformations. In doing this, these chapters continue the reconstruction of Konso history, from the 1950s to the present day. Through this history, the relationships between culture, landscape, individual livelihoods, and Konso institutions, are further examined.

The value of Bourdieu's (1977) model becomes evident here. The processes that are described in Chapters Five and Six represent, in Bourdieu's terms, the heterodoxy. They represent sets of modern practices, ideals and values that challenge those of the orthodoxy described in Chapters Two, Three, and Four. This heterodoxy has challenged the orthodoxy, sometimes almost to the point of annihilation. But, the orthodoxy is more resilient than might have been expected, and the two discourses also exist contemporaneously in debate with each other. This debate – or dialogue – between the customary and the new, the orthodoxy and the heterodoxy, changes over time, with the relative strengths of the different discourses and positions.

Chapters Five and Six explore processes that are described here as forms of modernization, as they are processes through which aspects of modernity have impacted on Konso, and developed in it. Chapter Five explores the new religious beliefs and practices which arrived in Konso in the middle of the twentieth century. These new practices rivalled and undermined the religious and ritual roles of the *poqalla*. The discourses of the new religion recast the *poqallas*, and the practices they embody, according to their own readings of Konso culture, imposing their own valorized and binary views of the world. While their own beliefs and practices are constructed as 'Godly', those of the *poqalla* are seen as being of the Devil. As, in the institution of *poqalla*, the practical management of land and labour is integrated with the ritual ideas, beliefs and practices, the destruction of those ritual ideas, beliefs and practices has had far-reaching consequences for the management of land and labour. The material in this chapter illustrates further the major significance of religious ideas and practices to the practical matters of agricultural management, land tenure, indigenous knowledge and landscape.

The religious changes described in this chapter have not taken place in isolation. From 1974 onwards, Konso experienced dramatic changes following the coming to power of the *Derg* revolutionary government. The ideologies promoted by scientific socialism combined with the new Christian ideals to even more powerful effect. Chapter Five describes the history generally and focuses on religious change and experience. Chapter Six focuses on the changes brought about by the revolutionary *Derg* government. Chapter Six also reviews the way in which the changes brought by church and state have intersected. In addition, the impact of these two interconnected processes has not gone wholly unresisted by Konso people. The way the different groups of Konso have accepted and rejected different aspects of these processes over time, is also discussed in these chapters.

Chapters Five and Six explore changes that have taken place over the last century that have radically transformed Konso people's senses of themselves and their place in the world. For example, when Konso people were incorporated into the Ethiopian Empire at the end of the nineteenth century, they became part of a wider state project. There were new kinds of information and new demands from the newly formed capital, Addis Ababa. As the twentieth century unfolded, the demands and interventions of the state increased. When the foreign missionaries came to Konso they incorporated Konso into a network of contacts that stretched further afield. The missionaries and the state together brought education, literacy, modern medicines, new technologies and forms of transportation. Even if all could not benefit from these developments, they became exposed to and aware of them. They changed people's ideas of who they were and of what was possible.

Donham (1999) describes similar processes in Maale. He argues that modern education, new technologies and markets combined with the values and ideologies of those who brought them, and together these created a modernity in Maale. This modernity radically reordered Maale understandings of themselves and their place in the world. They changed Maale imaginary geographies and their imaginary histories through which people understood who they were, where they were going, and where they had come from. The changes also brought new meanings and values which were attached to these ideas of self, place and world, and to particular lifestyles.

The modernity that was produced in Maale was a vernacular modernity: in other words, it was locally constructed and it had a local form (see also Miller, 1994; Ferguson, 1999). But it can be described as 'modern' because it shared certain characteristics with other 'meta-narratives of modernity' that existed elsewhere. Most particularly, it employed an evolutionary framework which 'ordered nations as being either "in front" or "behind" one another, all on a straight line, defined by relative wealth, power, and knowledge' (Donham, 1999: 2). It valued certain kinds of lifestyles above others. This 'myth of modernity' originated elsewhere. It had roots in Enlightenment thinking and was developed as a strong discourse in academic and popular literature. It also developed as a set of assumptions that were held deeply by governments and policy makers. But although external in origin, these assumptions and discourses were internalized and reproduced at the local level in powerful forms, to the extent that even when these ideas were discredited in international debates, they were still frequently employed at the local level in Africa, and other developing areas. As Ferguson has shown, they were perpetuated in 'clichéd dualistic stereotypes', so that a conceptual framework that ordered people into different stages on an evolutionary scale, with traditional at one end and modern at the other, became in many cases, 'a local tongue' (Ferguson, 1999: 84).

A second aspect of the meta-narrative of modernity has had a powerful impact in Konso, but is less discussed in the literature on change and transformation. It relates to the optimism and sense of empowerment that is associated with modernity. This optimism is derived from the Enlightenment beliefs in science, reason and progress. In the modern discourse, science is seen as having the capacity to improve lives, and to free people from the vagaries of nature and disease. Science and reason would facilitate 'progress', an 'ever increasing well-being and happiness, a movement from "badness" to "goodness"' (Schech and Haggis, 2000: 4). With modernity comes the belief and expectation that it is possible to transform the world and to take control of fate. It is possible to move on to an improved, and possibly ever-improving, condition of life. In Konso, where the environment is capricious, and disease, hunger and misfortune can strike at any time, this aspect of modernity made this set of ideas persuasive.

In order to understand the changes that have been brought about by church (this Chapter) and the state (Chapter Six), it is important first to set out the broader context and the main developments in Konso over the twentieth century.

Historical context of modernity in Konso

Some of the history of the incorporation of the south into the northern Empire has been described in the previous chapter. There, the role of *poqallas* as intermediaries with the state was described. Here, I lay out the history more generally, and explore the attitudes towards the incorporation of Konso into the northern Empire, and the way it was experienced. Other works that explore in detail the experiences of southern peoples of northern rule include Donham and James (1986), Markakis (1974), Bahru, (1991).

Konso was brought under the rule of the Abyssinian Empire under Menelik II in 1897. The northerners set up headquarters in Gardula (just below Gidole) and they had a base in Chencha. There is also an area in the lowlands on the road from Konso to Gummaide where it is said that the Governor, Habta Giyorgis, had a house, and Konso people often refer to this period of northern rule as 'the time of Habta Giyorgis', or 'the time of Lulsaggad'. The former is probably a reference to *Fitawrari* Habta Giyorgis[1] who received the Governorship of Borana in the extreme south, and who later became Minister for War (1907–26) (Bahru, 1991: 88). The latter is probably a reference to the military officer *Dajjach* Lulsaggad Atnafsaggad, who played a key role in many of the power struggles of this period, and who later became *Ras* Lulsaggad, killed in 1916 (ibid.: 128).

Oral histories of this period in Konso tend to stress the often contradictory experience of education, modernization and subjugation which was the nature of their inclusion into the Ethiopian Empire. As one Konso man, Kalzimo Dinote, explained:

> When Lulsaggad came the people didn't know anything. Lulsaggad was the leader of the northern people. They came to teach the people, but the people did not like them and they fled from them. So the northerners forced people to live with them by using the gun. The people saw the gun as being like the *okalla* container, which is filled with ash - when it is hit all the dust puffs out. The northerners told the people to pay tribute but the people said, what are these people and why are they asking for this tribute (*gabbar*)? The people turned more against the northerners and the northerners began to shoot them, and the people saw the northerners filled the *okalla* [a skin container] with ash and they saw that it killed people, and all the people continued to run away. The northerners continued to ask for tribute and many resisted and were killed.
>
> Then the people said '*holla! holla!*' which means this thing is more powerful than us and we can do nothing, so we are out of this thing, and they stopped resisting. Those who still resisted were shot. Then the northerners warned the people not to resist and they put their representatives in the area, who were called *melkagna*.[2]

The Konso submitted because of the superior arms of the northern soldiers. Even today, the Konso word for northerners, '*Qaweta*', is derived from the Oromo term *kawwe*, which means rifle. Other interviews in the field related how Jarso and Dokato were the main villages that resisted the northern invasion and received the brunt of the northern attack. Other villages, on seeing the fate of Jarso and Dokato, eventually submitted to the new regime. The northerners were based at Dokato and Bekawle and also at Fasha and Turo.

The northerners demanded tribute in animals, in produce and in labour, which was known as *gabbar*. Those who could not pay tribute were imprisoned. For example, as Kalzimo continued:

> The *melkagna* forced people to work for him. Each family chosen to work for him had to send one member of the family at the appropriate time, and each family worked equally, it didn't depend on the size of the family, whether there were three or fifteen members, they all worked equally.

> The people who worked for the *melkagna* were simply selected and forced to work. If they didn't work then they were beaten. My father was chosen to be *gabbar* simply at a meeting. If he had tried to leave he would have been killed. Other people were captured on the road for the work. He built houses, fetched water and kept animals. Twenty families were assigned to each *melkagna*. In one week some would work, and in another, others would.
>
> Chencha, Gidole, Gommaide, Iavello, all had representatives of Habta Giyorgis and Lulsaggad. The *melkagnas* had their own police they called *necholabash* [meaning, wearing white]. Each *melkagna* had about three *necholobash*. There were also people who were clerks to do the reading and writing for them. There were northerners, living in Karate, Fasha, Turo, and Gawwada.[3]

The newcoming 'northerners' included soldiers, administrators, clerks, court officials, police, and their families. There were also priests, traders, entrepreneurs, hunters and adventurers. Although many Konso people may have been impressed by the trappings of Empire, and by the exposure to the courts with their scribes and so on, in the main it was seen as an occupation that was simply made possible by superior force.

In 1936, the Italians took control over the country and ruled until 1941. Existing literature does not explore in detail the period of Italian rule in the south. According to a northerner who came to Konso with the expansion of the Empire, and who was still living in the 1990s, the Konso resisted the Italian invasion at first, but they were attacked on the ground and from the air, and were soon overpowered.[4] In Buso, the people still talk about the way in which the Italians shelled the village of Xormali with a big gun from Shanqella, a hill above Buso a distance of approximately 3km. The people had never seen such large artillery in the region. The villages of Aba Roba, Gaho and Fasha were also severely attacked.

In Buso, Kahano Kalayta and Kahano Kolobo explained that the people saw the suffering of other villages and so they gave the Italians food, *chagga* (beer) and honey, so that they would not be attacked. They explained that the people had become used to rule by outsiders, and that 'they did not mind the change in rulers very much, they [all the rulers] just had guns and therefore power. We haven't fighting instruments, so we have to be with the people in power'.[5]

Amborn (1988) writes that because the Italians abolished the system of *gabbar* labour, and because the rule of the northerners was resented so much, the Italians were received as liberators in some areas of the south. Kembo, the Buso man who was imprisoned by the northerners when he was a young boy (see Chapter Four), and who feared and hated them, confirmed this when he described his own emotions when the Italians came:

> After I was released, my elder brother was caught by the northerners. Then the information started to come: 'the Italians are coming, the Italians are coming.' And we asked 'Where are they coming from? What will they do?

> Are they really coming?' We thought that they will destroy these landlords, and that this is good. After two days they arrived in the administrative headquarters, and they killed some people and they burned some houses. When we learned this we were so happy that we sang and danced in the village clearings *(mora)*.[6]

The Italian invasion was not viewed in such positive terms by everyone, and during this period, the community was split. Those who resisted the Italian regime and who were captured were sent to Gidole where many of them were killed. Others fled to surrounding areas, where they joined with northerners who were leading a guerrilla campaign against the Italians. They formed groups of resistance fighters known as *'Banda'*.

Others cooperated with the Italians, and received a valuable wage for doing so. Kembo again explains:

> Our enemies [the northerners] were overthrown.[7] The people came close to the Italians and cooperated with them. Thirty people from Buso went voluntarily to fight with them. They went to Hamar [to the south-west] and fought there, and my father died in this battle. Those who went were paid according to their position: some 60, some 100 *birr*. Then the money was called *'hanchiteta'* – each is worth 10 *birr*. Some would take 10 of these, others 6.[8]

But many continued to resent the control by outsiders. It is generally discussed as yet one more time when the submission to a superior force was unavoidable; the rule from outside was still viewed as extractive.

After the Italians were defeated in 1941, the northerners led by Haile Selassie continued a concerted and ambitious programme of modernization of the country. There was a new emphasis on education and health development. Salaried positions were established in the provinces to replace the *gabbar* system and to undercut the semi-autonomy enjoyed by the regional governors who had previously relied on receiving a share of the tribute payments. Markakis writes of the reasons behind these developments

> It was gradually realised that the multiple exactions imposed on the *gabbars* by the landholders constituted a heavy burden on the peasantry, without, moreover, adding anything to the state treasury. Furthermore, the landholders themselves were largely exempted from taxation (Markakis, 1974: 118).

This period in Konso memory is dominated by one man, *Fitawrari* Tadessa Wolde, who lived in Dokato and who governed Konso on behalf of the Empire for over thirty years. He managed, to some extent, to resist the new limitations on his power and position imposed by the modernizing state. Donham describes how the average time in office in Maale was between three and six years, which shows that Tadessa Wolde had a remarkably long tenure in office.[9] One Lutheran missionary, who worked in Konso at the time of Tadessa Wolde, said that Tadessa Wolde had his position terminated by Addis Ababa many times. Each time, he was reinstated, because he was married to someone who had powerful relations at court.[10] He was a man of powerful stature and personality, and it is said that he spoke *Konsiña.*

Tadessa Wolde appears to have been able to continue to live as many northerners in Konso had done before. He continued to take taxes and also labour tribute from the people. The women still had to grind teff and *berbere* and collect firewood and water for him. As a group of women explained to me:

> Before, at the time of *Fitawrari* Wolde, when you went to the fields to collect grass and leaves for the animals, the people [northerners] would catch you and take your grass and leaves. When they saw you, they caught you and they took you to go and grind their teff. Sometimes they said that it was not done finely enough, and they would say 'grind it again!' Even then they could mix the ground grain with fresh grain so that you would have to do it a third time.[11]

At this time, the central state set up a complex system of taxation. The first land tax was introduced, which levied charges depending on the quality of the land that was farmed, classified as fertile, medium or poor. In addition to the land tax, a tithe was also payable, which was one-tenth of the annual produce of land or labour. In 1942, these taxes were regularized and payment had to be made in money. In 1947, an education tax was also introduced of between 4.50 and 15 Ethiopian dollars, and in 1959 a health tax of between 4.50 and 13.50 Ethiopian dollars (both depending on whether the farmer was classified as using fertile, medium or poor land) (Markakis, 1974, 121). These figures are for Gemu Gofa region of which Konso was a part. Thus in Konso this period was characterized by the double burden of a continued extraction of labour tribute under Tadesse Wolde, and a gradual increase in levels of taxation. This situation continued until the overthrow of the Empire in the revolution in 1974.

Christianity and conversion

The previous chapters have shown how the practical aspects of people's lives were bound up with the social and ritual dimensions. Thus, the arrival in Konso of new religious movements wrought profound effects. Until the last ten years or so, the influence of Islam has been very small, but the Ethiopian Orthodox Church and the Protestant Church have both played a part in people's lives, although their influences have waxed and waned over time.

Ethiopian Orthodox Churches were built in Fasha and Turo during the first period under the northern Empire (1897–1935). During the second period, a church was built in Bekawle, now Karate, the main town in Konso. These churches served mainly the northern population, but also some Konso. It is difficult to estimate the numbers of Konso people who converted to Orthodox Christianity at this time, but it seems that it was not many. It has already been described how some of those who co-operated closely with the northern administrators converted, including the regional *poqallas*, Bamalle and Kalla. Some others complained that they were made to convert by force, as the following account from one Orthodox Christian explains:

> During the time of Haile Selassie, people had their heads shaved and they were baptised by force, and the people hated this. The saying was, '*kirarayiso Wagiteta an ingito*' – 'to Christ by force cannot be made'. People said, 'I was made a Christian by force and so it cannot affect me.'[12]

The speaker described how such forced conversions were resented, and cannot be considered to be 'true' or lasting. In the mid-1990s, the number of Konso people who considered themselves to be Orthodox Christians was relatively few. At the same time, there were many Protestant Christians.

The most important Protestant Christian church has been the Ethiopian Evangelical Church Mekane Yesus (EECMY, henceforth described simply as *Mekana Yesus*). It has worked there since 1954 in conjunction with Icelandic Missionaries. Konso was, in fact, the first mission station of the Icelandic Missionary Society in Ethiopia, although it worked together with the more established Norwegian Lutheran Mission, which had a mission station in Gidole. Together, these churches promoted a Protestant Christian faith which had a profound effect on many in Konso.

In the early years of the Ethiopian Empire, the work of Protestant Christians was discouraged, as it was seen as a threat to the long established Ethiopian Christian Orthodox Church (Fargher, 1996; Donham, 1999; Aren, 1978; 1999; Eide, 2000). From the 1920s, however, Haile Selassie, encouraged the work of Protestant missionaries, whom he saw as bringing valuable resources and skills. Under Haile Selassie their proselytism was forbidden in the north of the country where the Orthodox Christian church was strong, but was permitted in the south, where the people were viewed as 'pagan'.

More particularly, the missionaries' construction of schools and health centres was encouraged, and their deployment of doctors, nurses, construction workers and teachers was considered valuable. The missionaries were encouraged because they became another means through which Haile Selassie could achieve his aim of modernizing the country. As Donham writes:

> After the Italian occupation, Haile Selassie's government stepped up its pressure on evangelical missionaries to provide modern services – particularly schools in the late 1940s and 1950s… (1999: 104).

> [The Emperor's] own modernist goal was, after all, to wring as many hospitals and schools as possible from the missionaries - not to mention threshing machines and Ford automobiles (ibid.: 94).

Donham also quotes Fargher, who discussed the way in which the Empire encouraged missionaries because of the role they could play in the modernization process:

> The government needed pliable change agents. Who better than expatriates that had no political ambitions, who believed in the saga of a Solomonic dynasty, who were willing to educate and carry on medical work? They required no money... The government obtained an influence from the expatriate evangelists: the influence of modernity (Fargher, 1996: 995–6; in Donham, 1999: 94).

In Konso the missionaries built a mission compound near to Bekawle, then the headquarters of Konso. The compound was large and contained a church and houses in which the missionaries and others from the Protestant Christian community lived. Wilfred Thesiger camped near the mission compound in 1963, when he was travelling through Ethiopia:

> We camped near an Icelandic mission at Balalawi [Bekawle] ... These particular missionaries happened to be simple, friendly people; nonetheless I did resent their presence. How many missionaries, I wondered gloomily, representing rival denominations from Catholics to Nonconformists, Baptists, Seventh Day Adventists and others, were now at work in Southern Abyssinia, gratuitously disrupting the life of these tribal people? (Thesiger, 1987: 408).

Beside the church compound, the missionaries built a primary school and a health centre, both of which are still functioning today.

The history of one Protestant Christian convert and his family in Konso can be related to reveal the nature of conversion and the struggles it entailed. It also demonstrates the way in which the spiritual dimensions and powers of the new religion were inseparable from the resources and skills that the mission churches brought with them and offered to the people. The following account is from Estiphanos Berishia, whose mother and father were among the earliest converts to Protestant Christianity. At the time of the interview, he was training to become a pastor. Because of his education, the interview took place in English. The story is of particular relevance to this study because Estiphanos' grandfather was a very powerful *poqalla*, and, as the eldest son of an eldest son, he himself would have inherited the position of *poqalla* in due course. The story gives an account of a *poqalla*'s life, and the way it changed when faced with alternative ritual and religious ideas and practices. The first part of his account gives some general details which set the scene for the conversion of his mother, his father, himself and his sister:

> In the past, my family was a *poqalla* family and they were very important. We had many wide fields and many of these fields were given to other families. We gave fields to around forty families. When my family needed help, for example to dig the fields, these families came to help them, because they respected them. At times, for example, when my mother gave

birth to a child, they used to collect money and contribute and just present their gift. My grandfather also collected coffee from all of these fields.

Our family was also *balabbat* at times, but not always. But my family was important: one, because of its wide fields; and two, because my grandfather was a leader in the traditional religion. For example, when there was dancing, he was the leader. During rituals he was the leader. And he was the leader of all clan members.

In 1968[13] my father and mother became Protestant Christians. Why? Because I was sick. It was said that it is because my grandfather took money from a woman and didn't want to pay her back, and so the woman went to a witch and 'sold' us as we say in Konso – *setana ekani*. She accused us and said that *setana* [a spirit] must punish us by killing our children.

At the time my mother was already old. She had spent eighteen or seventeen years as an adult before she was married. She was probably close to forty years old when she married.[14]

This discussion of Estiphanos' mother's age refers to the first generation-set in Konso called *farayta*, which features prominently in many Protestant Christian narratives. When a person is 'in *farayta*' it is forbidden for him or her to have children. Because entering a generation-set in Konso depends on the generation-set that a person's father was in (rather than a person's age), it is possible for a person to reach an age between thirty or forty before he or she finishes *farayta* and is able to have children.[15] A person is permitted to marry when in *farayta*, but it is thought impossible for someone in *farayta* to have children (Hallpike, 1972: 141). In Karate region, if a woman became pregnant during *farayta* then it was thought that she should have an abortion. In Fasha, the pregnant woman would have to leave her home village and remain in what is described as 'another country' – a liminal space – until the years of her time in *farayta* were over. When this period was over, the woman and child would ritually re-enter the village. They would be covered by an animal hide which would be removed on entering, as if the child were being born for the first time. *Farayta* is not kept any longer because of the way in which the practice was targeted by the missionaries and, as will also be seen, by the revolutionaries. However, it accounts for the age of Estiphanos' mother, as she may not have been permitted to conceive until late in life. Estiphanos continued his narrative:

After she married, my mother gave birth to twins. Then, my mother saw in her dream the woman who had accused my grandfather. She came to her at night in a dream and took the baby from her breast where the baby lay and was sleeping. In the morning my mother saw that the baby was very sick, and within a month both of the babies died.

After I was born, I was sick also. And this woman was still insulting our family. My mother went to see Sangogo [a *poqalla*] who was a *sawaita* [has powers to see into the future] and he told my mother that unless she becomes a Christian I would die also.

> Sangogo was not a Christian [at that time], but he sent her to another Christian to be advised. This man said to her 'you are a wife of a big and famous family. If you become a Christian you will be persecuted and you will never be in this big house with all of your luxuries. Your name will be spoilt, you will lose your reputation and everyone will be against you, but your son will be saved.' So my mother decided 'Let people come even now and slaughter me, and kill me, if only my son lives, I will be a Christian.' So she immediately woke me and my younger sister and took us to church. We went there and later my father also became Christian.[16]

In explaining his father's conversion, he said:

> I am the only son for them, and as you know in Konso, if people do not have a son who will inherit the possessions, we say that he is not alive anymore, because they believe that they live through their descendants, through their seed. If one does not have a son then it means his life is cut there and it will not continue any more. This is very bad. We say that people who get like this are the worst sinners among people. That's why God stops their life there and doesn't give them any continuing generation.[17]

In this account it can be seen that, even in the case of a man who has been formally educated and brought up as a Protestant Christian in close proximity to the mission, the motivation for becoming Christian is influenced by a combination of Protestant Christian ideas and indigenous ideas concerning wellbeing, fertility and disaster. The indigenous model of the worst case scenario is one in which people do not have children, or where those children do not prosper. As Chapter Three has discussed in more detail, disaster is thought to result when people do not respect the culture (*aatta*), and when people do not honour their *poqalla,* or when they transgress in one way or another. The Protestant Christian teachings fit rather neatly into this framework: the transgressions are understood as sins which are punished. The power of the Protestant Christian teaching is that it provides a way for these sins to be atoned for and forgiven, so that the disastrous consequences can be avoided.

In Estiphanos' telling of his own story, Biblical ideas and images can also be seen to have influenced his understanding of what happened. His mother's brave stance, and his father's ordeal (described below), resonate with Biblical tales. The struggles against adversity that feature prominently in many Protestant Christian narratives may have coloured his account. But despite this, his story provides us with insights into an early case of conversion to Protestant Christianity, and a deeper understanding of how that conversion was thought about and experienced.

Following his mother and father's conversion (and that of himself and his sister) his father experienced some problems:

> At one time, people tore the clothes of my grandfather and took him to the court. And they accused my father of attacking him [the grandfather], and said that my father was about to kill him. So the police came at once and

> captured my father and they put him in jail. At that time Mr Gisli [an Icelander] was the missionary here, and he ran to Arba Minch and to Addis Ababa to tell the government that the local people here and the administrators are mistreating Christians. It was not only my father but others were also treated in the same way. Then the judge called my grandfather and my father, and my grandfather said 'I did this accusation against him because he left our traditional way of life and became a *ferenji* [white man]. He accepted what these white people brought to us, so it is because of this that I did this reckless thing.' So the judge said that he had done wrong and he set my father free. But when my father was freed, many people came on the road who were ready to kill him. And so *Qes* Berishia [a Konso man who had become a Christian and a priest] and Mr Gisli came with the policemen and rescued him and took him to *Qes* Berishia's house, where he spent the night.[18]

Estiphanos also talked about his own experience following conversion and the important support that he received from other Konso Protestant Christians:

> At that time [of conversion], I was a very small child and I was suffering from tuberculosis. They immediately put me in school and I started my schooling. Later, I was attending school and at the same time suffering from TB. At that time, there was a very good Christian life and fellowship and they were helping us very much. They were paying the money needed for my treatments, and at times when I was put in the health centre they also paid for this. In 1970, I was brought to Gidole and was there for almost a month [in hospital] and the church also assigned people who would come with my father and my mother to be there with us.[19]

When Estiphanos' mother and father converted, therefore, their lives and those of their children, were 'saved' in more ways than one. They were 'saved' in the sense that the word is used in evangelical churches to mean leaving behind a life of sin, and joining a community of Protestant Christians who hope ultimately to go to heaven. But they were also 'saved' in the sense that joining the Protestant Christian community brought access to modern medicines, hospitals or health centres, and to the financial and personal support needed to overcome the illness. In a place where modern medicines were previously unavailable, it is impossible to overestimate the power that the missionaries held through their ability to provide these resources.

The connection between modern medicine, missionary activities and conversion experiences is familiar from other parts of Ethiopia and Africa more generally. It features prominently in Donham's account of the work of the Sudan Interior Mission (SIM) and conversion in Maale (Donham: 1999). Comaroff and Comaroff (1997) also examine the way in which medical healing and spiritual healing were both seen as part of the responsibilities of missionaries in southern Africa. Whether or not the missionaries saw their medical and educational responsibilities as part of their work of saving souls in Africa (as the Comaroffs argue in the Southern African case), or as material matters that potentially distracted new converts from what they saw as the real meanings of Protestant Christianity and the real reasons for conversion (as Donham describes in Ethiopia), matters little.[20] In practice, the combination of prosyletism and modern medicine was inordinately powerful.

The connection between the provision of medicine and Protestant Christianity has continued. The status of the mission health centre is currently uncertain, but while it was working regularly, it did not restrict provision of services to Protestant Christians. Nevertheless, in the wider community the impression persisted that it did. At the health centre all patients joined in prayer at least once a day, and although participation was not forced, non-participation was frowned upon. When the mother-and-baby clinic took place, the young (Protestant Christian) nurses took the opportunity to talk to them about Jesus and to show them the Bible.

Modern medicine is not the only service that has been provided by the Protestant Christian church. The first schools in the area were also established by the church, and new Christians such as Estiphanos were among new pupils encouraged to attend. They received the help and support necessary to continue their education. Education was by no means restricted to Protestant Christian children; but conversion to Protestant Christianity often followed school attendance, because of the influence of teachers and peers. The connection between conversion to Protestant Christianity and school education also persisted in the state-run schools which were established later. It led one woman, for example, whose eldest son would in time inherit the position of *poqalla* from his father, to refuse to let her eldest son go to school. She argued that it would lead him to become Protestant Christian and 'leave the culture'.[21]

From this account, it is possible to understand the main pull-factors for conversion to Protestant Christianity in Konso. Protestant Christian ideas and indigenous ideas about transgression, sin and disaster fitted conveniently together. The claim of the Protestant Christian teachings to be able to 'save' a person was taken quite literally and made the religion attractive. In practical terms, Protestant Christians brought with them the Konso people's first access to modern medicines, and made the 'saving' of lives a reality. The missionaries also organized and carried out food distributions and small-scale development projects on behalf of the church. They provided education and new opportunities. They also had cars and provided transport to those who were in a position to ask for a lift to the nearest town. They had contacts in the rest of Ethiopia and wider afield, which could be drawn upon for support, further education, and employment opportunities.

Joining the *Mekana Yesus* church brought with it practical, social, and sometimes even financial support from a vibrant community. A convert became part of a network of contacts, support and opportunities, not to mention the books and stories that the missionaries and other Protestant

Christians brought with them. These were novel pleasures for people in a place where books and radios were almost unknown.

Other work on conversion has commented on a further motive: conversion is attractive for many because it allows them to step out of many of the norms and constraints to which they must conform in their 'traditional' cultures. Thus, it has been argued, conversion is particularly attractive to women, who find relative freedom in the church community. The enthusiasm that many young people have shown towards conversion, and the wholehearted commitment to church activities they display, suggests that there is some truth to this in Konso. In church, young people can meet, sing and dance with other young people in an environment where such behaviour is considered appropriate and legitimate. Among the more charismatic groups in the church, the free expression of emotion and passion for the religion is even considered praiseworthy. These opportunities for socializing are not wholly new: young women and men always have had the opportunity to meet each other and to spend time with each other, for example in work groups. In rituals (for example, weddings, funerals,[22] or the feasts that accompany the end of the cultivation season), young people are permitted to dance and sing together. Nevertheless, the church may provide opportunities for socializing, and for people to take part in new forms of dancing and singing; men appear to be as enthusiastic as women about joining in.

The existence of these practical, material and lifestyle advantages to conversion does not mean that the belief in Christianity should necessarily be seen as shallow, suspect or disingenuous. Christianity is a powerful religion. The spiritual, psychological, and experiential dimensions are meaningful and inspiring for many Konso people who convert. But the indigenous context has influenced the local form that Protestant Christian beliefs and practices have taken. Those Protestant Christian beliefs have also reinterpreted the indigenous ideas and beliefs in ways which have provided powerful pull-factors for conversion. The hybrid of Konso ideas and Protestant Christian beliefs, as held in Konso, proved to be particularly bad news for the *poqallas*. Conversion to Protestant Christianity, and joining the *Mekana Yesus* church, also involved a simultaneous rejection of all things considered 'cultural'. In order to understand this process, it is necessary to turn to Konso ideas about spirits and spirit possession, and to the way they were viewed and debated by Protestant Christians.

Spirits, spirit possession and Protestant Christianity

The word *setana* is used in Estiphanos' story to refer to a woman who 'sold' another family by going to a witch and paying for misfortunes to be visited upon them.[23] The reference to *setana* demonstrates that an individual's mistakes are not the only reason believed to be responsible for misfortune in Konso (and hence for conversion). The harmful intentions of others and harmful external forces can also be held to be responsible for disaster. These forces are known in Konso as *ayana*, or, as they are described above, *setana*, a word that gains additional meanings because of its similarity to the word Satan.[24] The Protestant Christians have interpreted and re-interpreted the concepts of *ayana* and *setana* in a way which casts a particularly negative light on Konso culture generally, and on the role and position of the *poqalla* in particular. The concept of *ayana* or *setana* has caused the debate about the relative positions of Konso culture and Protestant Christianity to become polarized and extreme.

Setana, or to use the less inflammatory term, *ayana*, is used to refer to supernatural forces, usually conceived of in terms of spirits, that have the power to act on a person resulting in either wellbeing or disaster. Debates about *ayana* provide another explanatory framework for the fertility, health and happiness that one household experiences, and the illness, deaths, and misery of another. This framework is different from the one that refers to 'correct behaviour' and to the *poqalla* as described in Chapter Three.

The Konso ideas of *ayana* and *setana* refer to spirits which are extremely powerful, and have the ability to bring great fortune, wealth and power to a family. But they also have the ability to wreak havoc and bring disaster on a massive scale. The nature of *ayana* is best explained through Kahano Kalayta and Kahano Kolobo's words, and here, as elsewhere, the words *ayana* and *setana* are used interchangeably:

> Those spirits of *setana* that come to people come from nearby those lakes [indicating the Rift Valley lakes, for example, Lake Chamo, to the north of Konso]. They cheat some families and make other families rich. For some there is the birth of a lot of children and the containers burst accidentally [the harvest is so large that it cannot be contained]. The grain ripens, the sheep get fat. Everything in the house is fruitful, the beehives give a lot of honey. Everything becomes abundant. Then after all of this abundance, it begins to disturb the family, and it catches one of the children or the father of the family and it starts to make them sick. And it says 'why don't you tie something for me, why don't you tie a sheep or a goat?' It takes revenge. If they then tie something, then it releases the person whose blood was being drunk, and it goes to drink from the animal that is being tied.
>
> Normal sheep don't grow so much but the *ayana* sheep that are tied are bought when they are very small and then just within one or two years '*Posh*!' they have grown up to be about the size of a bull. Then the *ayana* demands people to slaughter the animal. The *ayana* says, 'Give me my yearly blood', and the family then searches for any animal that resembles the former one in every feature... In this way all the property of the house is finished. People will even sell a piece of land to buy an animal.
>
> If any mistake in this culture is made, then it will kill a child of the house or the mother of the house: for example, if they bought a goat with a slight mark of red or white on the belly, or a slight cut on the ear. Then the *ayana* comes and boasts about the killing and says 'I killed them, I told you how

to do this work correctly and you made a mistake, so I killed them.' This is why people hate *setana* and they say they don't want it.

This is the story, so let us leave it now.[25]

The last sentence reflects the reluctance of people to talk about these matters today, as many find even talking about them threatening.

Ayana can benefit a person. It is said to make him or her lucky; everything he or she touches becomes abundant and he or she becomes wealthy. The abundance experienced is beyond what is 'natural'. Through *ayana*, a person can remove himself or herself from the conditions within which he or she lives. *Ayana* is empowering in that it gives someone the ability to 'cheat' the constraints of the environment and his or her own innate capabilities: the fields give huge amounts of grain and the family prospers. One man told me how, in the 1960s, he had been possessed by *ayana*, and he had been so strong and powerful that he had married nine wives. (To support nine wives he would have needed significant economic resources, but his marriages were presented to me as a reflection of his powers to make anyone he wanted marry him.)

But such powers are risky and have a cost. A person may benefit beyond his or her wildest dreams, but they also suffer from the 'unnaturalness' of what they have been doing. Some people become possessed by the spirits with whom they have been associating, and they obtain superhuman strength. As a result, their behaviour becomes erratic and dangerous: they are said to have a tendency to leap around in trees or to climb into others' compounds and disturb people. In dire cases they can throw themselves over cliffs. One man I knew well showed me the scars on his arms where, he said, he had been tied by the village in order to stop him running around and causing damage to himself and others, when he had been 'strong' with *ayana*.

Most terribly, *ayana* can come and 'sit' on a person and make him or her mad, or thin, and eventually cause that person to die. This is evident in the quote above, where the capricious spirit becomes tired of obtaining things for the family, and starts to demand more and more sources of meat, blood and fat, to satisfy its own insatiable hunger. In order to appease the spirit, the family must sacrifice animals as the spirit demands. As the spirit particularly likes to consume fat, the fat of the animal is smeared on to the central post of the house, and the spirit can eat it there at its leisure. If the spirit's hunger is not satisfied (and its greed is ever increasing), it is said that the spirit will come and 'drink' and 'eat' from one member of the family, who weakens and finally dies. *Ayana* is thus thought to be greedy, malicious, fickle and unforgiving, and has the potential to kill everyone in a household.

In some families, the good period of wealth acquisition is not experienced, and they only suffer at the hands of a spirit. Particularly malevolent spirits can also cause destruction on a massive scale: one whole neighbourhood in Buso village experienced many deaths over a short period in the middle of the last century. Although probably the result of a communicable disease, it was, and still is, understood to have resulted from the actions of a powerful spirit that lived nearby. In some cases, spirits are supposed to have been sent by someone who wishes harm (as described in Estiphanos' case above). In other cases, becoming attacked by a spirit is simply thought of as being a result of 'bad luck'.

Although harmful, capricious and greedy, the spirits are also admired by some because of their power and ability to bring wealth. The combination of fear and respect is evident in the quote above. Such ambivalent feelings towards spirits and witch activities have been described elsewhere in Africa and can help with the analysis here (especially Geschiere, 1997; Comaroff and Comaroff, 1993; Masquelier, 1993; Behrend and Luig, 1999). Geschiere's discussion of *ekong* witches in Duala, Cameroon, for example, shows how they are highly dangerous (they are thought to keep zombified workers on secret plantations), but that they are also admired in some ways because they provide a means to wealth and success. Indigenous ideas about witchcraft and spirits such as these provide a framework and a forum through which ideas about power, wealth, madness and desire can be debated, discussed and explained. The *ayana* spirits represent in microcosm some of the more extreme possibilities of human action and their consequences. Ideas about *ayana* provide a model for explaining the relative prosperity and power of one person over another, and for debating the acceptability and legitimacy of resulting social difference. In these ideas, the conflicts that arise between a person's desires and his or her social responsibilities are expressed. In *ayana*, for example, a person's wildest dreams in terms of material wealth are pictured: the sheep grow to the size of bulls; the calabashes are so full of grain that they burst. But the discussions and debates about *ayana* construct such desires and expectations as selfish, and as setting a person apart from others. These desires are therefore constructed as 'unnatural' and asocial, and those who pursue them (presumably often at a cost to others) are shown, ultimately, to suffer the consequences.

Ayana is thus another institution. It is a set of moral beliefs and practices and is seen as a way of accessing wealth which is outside the socially sanctioned practices. Such socially sanctioned beliefs and practices are those that revolve around the *poqalla* and have been described previously in this book. The construction of *ayana* as dangerous and subversive is a way in which alternative forms of wealth acquisition, that is, wealth that is quickly won and not regulated by the *poqalla*, are discussed and socially sanctioned (with a greater or lesser degree of effectiveness). But there is still ambivalence towards *ayana*, and the attraction felt towards these spirits is not wholly dismissed by the sheer nastiness of its potential ill effects. The material rewards of *ayana* are still attractive, open to all, regardless of their wealth and position, and some might say worth the risk. It is possible that *ayana* might be particularly attractive to a person for whom the formal structures of power and wealth, that favour elder brothers and *poqalla*, offer little.[26]

There is also evidence to suggest that the attractive aspects of beliefs in spirits in Konso, combined with influences from ecstatic spirit possession movements from other parts of Ethiopia, developed into a set of ideas and practices which were much more coherent and organized in the middle of the last century. Accounts of the 1950s and 1960s describe how groups of people would meet and play drums and make offerings to the spirits. They would induce the spirits and speak to them through mediums who became possessed in these sessions. Reportedly, the most powerful spirits spoke in the Oromo language.

Such spirit movements have been common in many areas of Africa. One interpretation views them as a form of resistance, because, when possessed by a spirit, a person can make claims which otherwise would not be socially sanctioned. In this way, spirit possession can challenge the socio-political and moral order, while those involved can deny any agency, and blame the supernatural spirits (Comaroff and Comaroff, 1993). When the spirit speaks in a foreign language, its authority and foreignness are reinforced, and the distance between the spirit and the person possessed is increased. Spirit possession movements often intensify at times of extreme change and through periods of turmoil and insecurity, as individuals try to regain control over the world around them (Lewis, I. 1966; Lewis, H. 1984). It is for this reason that, Africa-wide, spirit possession seems not – as many would expect – to be decreasing. On the contrary it is reported to be increasing, and some have argued that it is a response to the forces and unsettling experiences of globalization (Comaroff and Comaroff, 1993; Geschiere, 1997; Behrend and Luig, 1999). Similarly, spirit possession activities have been identified among certain diaspora communities (for example see Boddy's work on Sudanese refugees in Canada: Boddy, 1995).

Spirit possession movements have been important in many areas of Ethiopia (Lewis, 1984). Many groups surrounding Konso, and particularly those with connections to the Oromo, have similar concepts that are even given the same or similar names. Knutsson (1967) describes how the Boran also have concepts of *ayana* and *setana*. Brøgger (1986) describes how the Sidamo (several hundred kilometres to the north-east) have *sheitanna* that they use to explain misfortune. His description of Sidamo ideas and practices applies equally well to the Konso: according to the Sidamo, a spirit announces its arrival to someone first with a sickness, but then the individual would manage his spirit, by giving it offerings at trance sessions called *hayatas* (Brøgger, 1986). The spirit would then assist that individual. The similarities between these spirits and ecstatic cults and those of Konso tend to confirm the argument that there was some external influence to the shape of the movement in Konso. The ideas of spirits were already present, but the formal meetings, the beating of drums and the speaking through mediums may have been incorporated from elsewhere. If it was borrowed from further east, then it is likely to have been fairly recent. Knutsson (1967) dates the origin of the movement in Boran to some time earlier in the twentieth century.[27]

When the missionaries arrived in Konso, they became aware of local ideas of *setana* and *ayana* and they featured strongly in their narratives about Protestant Christian teaching. The spirits were equated with the Devil, and spirit possession with Devil worship. The existence of the malevolent aspects of these practices was then used to legitimize missionary activities and to encourage conversion to Protestant Christianity. The negative side of spirits and the destruction that they were believed capable of causing was particularly emphasized, and the very real fear that they inspired was drawn upon to encourage conversion. One Protestant missionary, who was among the first to come and work in Konso, described her experience to me and how she and other missionaries perceived the situation of the Konso people (Baddingaldo is the name of the village neighbouring the mission station):

> When we first came here thirty-nine years ago, the people were walking in darkness and worshipping the Devil and he was like a great shadow in their lives. They were always afraid of him and he killed many people. Almost every night we could hear the drums in Baddingaldo, when people were worshipping the Devil. During this time, somebody would become possessed and put himself in the position of the Devil, and through this person the Devil would speak to the people. At night, people who were possessed would sometimes run around our house in the mission compound shouting and screaming, and it was terrifying.
>
> Many people came to us for help, to be saved from this and we told them simply to accept Jesus and let him into their lives. When they did that, their whole lives changed, from lives of fear to positive lives.[28]

Estiphanos – whose story is recounted above – described the situation in an essay he wrote on the history of Konso, as part of his training:

> The Devil's attack was not only during day times but also at nights. Immediately after sun set, the evil spirits appeared as bright lights moving out in the fields and flying like birds from one hill to another. Many people were possessed by them and became mad. Many committed suicide. Satan was very much feared, very much honoured, and was given offerings, much more than the existing government officers and administrators.
>
> The people were in this situation when the first missionaries arrived at Konso. The people quickly accepted the gospel, and churches were built in the villages. All things of Satan were destroyed. In the evenings, hymns were sung in every village. The bright lights disappeared. No more deaths, no more fears (Estiphanos Berisha, n.d.).

In interview he commented:

> Whenever we saw the Devil, we just quoted a verse from the Bible which we have studied by heart, that is, 'Hate Satan, and it will fly out, and come to Jesus and he will receive you.' We just recited this in Konso and the Devil went out.[29]

It is difficult to assess whether or not spirit possession was as common and virulent as these Protestant Christians portrayed it to be. It is possible that *setana* or *ayana* 'worship' intensified as a reaction to the processes of political, religious and economic change experienced at the hands of the northerners, Italians and missionaries, and that it came to a head in the 1950s and 1960s. This would be similar to the case in Sidamo, where Brøgger argues that Protestant Christians' teachings about spirits and spirit possession may also have given the spirit movements some more credence and power (1986: 70). It is also possible that it provided a set of practices and a network of people that was separate from that of the *poqallas*. In some ways, it might have been more democratic, in that membership and position in the networks did not depend on birth.

What is of particular significance here is that these ideas regarding *setana* and *ayana* reflected missionary attitudes to Konso culture in general, and to the *poqalla* in particular. *Mekana Yesus* proselytizers did not make a distinction between the ideas of *setana* and *ayana* and the practices and rituals of the *poqalla*. Instead some of the similarities were emphasized: the *poqalla*'s performance of rituals was seen in terms of performing dangerous magic. The spirits that visit the *poqallas* and who give them advice and can in some cases predict the future were interpreted as akin to *ayana* and *setana*, and thus as being from the Devil. In this way the work and lives of *poqallas*, and the ideas of *ayana* and *setana*, were conflated by the missionaries, and *poqallas* too were characterized as 'devil worshippers'. For example, Estiphanos' essay on the history of Protestant Christianity in Konso continues: 'Satan had chosen prominent families in the society and through this he ruled over the people' (Estiphanos Berishia, n.d., 12), where, by 'prominent families', he was referring to *poqalla*. Through associating the *poqalla* with *ayana* and *setana*, the *poqalla* and cultural life was demonized, or to use Meyer's term, 'diabolized' (Meyer, 1999).

The activities of the Protestant Christian missionaries brought a set of ideas and practices which reordered Konso understandings of their world in the same way that many of the modern discourses reordered the temporal consciousness in other places. In Konso, the missionaries and proselytizers had a hand in developing a discourse of modernity among the Konso, but the ideas of traditional and modern were polarized and valorized by Christian ideas of good and evil. Tradition, culture, *farayta*, *ayana*, the *poqalla*, and rituals were placed at one end of a linear scale, which represented the past, and was seen as a 'time of darkness' and suffering in which the Devil reigned. At the other end of the scale was Protestant Christianity, modern medicine and education, and a place and time in which there would be, as Estiphanos explained it, 'no more deaths and no more fears'. As the Icelandic missionary put it, 'whole lives changed, from lives of fear to positive lives.'[30]

The local socially sanctioned means of mediating against misfortune and disaster through the actions of the *poqallas* were tainted by their associations with the exaggerated narratives about *ayana* and other

Photo 5.1 *Two neighbours*

elements of culture that were viewed negatively. The abilities of the *poqallas* to protect and mediate against disasters were also outclassed by the promises of Protestant Christianity of salvation, combined with the power of modern medicines, education, and the other resources and contacts that the Protestant Christians brought with them.

The power of these associations was such that the community became polarized. There was little common ground between Protestant Christians and those who 'followed the culture'. Those who converted to Protestant Christianity eschewed many aspects of what they described as 'cultural matters', which they viewed as dangerous and connected to the Devil. This included attending rituals, dancing and singing in traditional fora, and in some cases, drinking *chagga*. As the *poqallas* were at the heart of the cultural and ritual life of traditional Konso, they were viewed as particularly dangerous.

This was particularly the case in the mid-1990s. Protestant Christianity was at its height, and was a powerful, even dominant force. Although converts to Protestant Christianity had been common and influential in previous decades (especially those who had completed secondary school), the activities of the Protestant Christian church had been curtailed in the *Derg* period (1974–91). In contrast, in the period that followed the change of government in 1991, there was a new momentum to the Protestant Christian Church activities. They were able to organize events, to evangelize widely, and their churches went from strength to strength. For example, new Protestant Christian churches were built in many villages, including Buso. There, young Christians were fired up by a new commitment. They constructed a church by raising money through forming Christian work groups (along the same lines as the *parka* or *marpara* work groups) and contributing the money they raised towards the church.

In general, at this time, it was easy to identify the most enthusiastic Protestant Christians. They separated themselves from others in the community through their activities, and also through their styles of dress and behaviour.[31] They socialized together and their activities were oriented around school and around the church. Where possible, they adopted Western-style dress (instead of the Konso *oowa* (skirt) or the colourful homespun shorts). Where this was not possible, they kept their clothes scrupulously clean. In many cases they embroidered their T-shirts with Christian symbols (see Photo 5.1), or decorated their houses with Christian symbols and sayings. Their language even was different, peppered with references to God. For example, the common greeting *'Nagaita?'* (Do you have peace?) usually responded to with simply *'Nagaita!'* was made into *'Nagaita Kalata Waqa!'* by the Christians, meaning 'Yes I have peace, thanks be to God'.

A brief comparison with Orthodox Christianity

This account of the impact of Christianity would not be complete without a discussion of the development and impact of the Ethiopian Orthodox Church in Konso. In the mid-1990s, the Ethiopian Orthodox Christian church was also active, but the number of its converts, and the scale of the activities of its members, was much less significant than those of the Protestant Church. The Orthodox Church was brought to Konso by the northerners, and the church was still strongly identified with them. Many northerners put down roots in the area, and their descendants still live there. Their first language may be Amharic or *Oromiffa*, but they also speak the Konso language and they are integrated well into Konso society. They may carry out some farming, but in the main they obtain a living from other sources, in particular trading. Some of them also have establishments that sell coffee, beer and food in the Konso towns. In recent years, some have also built small hotels.

As these northerners have been chiefly engaged in trade, they have formed an alliance with some of the Konso traders, who are members of the *xawuda* group. The *xawuda* group has an existing indigenous trade network known as the *fuld'o* network, which facilitates trade, sets conditions on trading behaviour, and lends money to its members. The combination of the northern trading network and the *xawuda* trading network has proved mutually beneficial for both groups, and provided new contacts, new information, and new trading opportunities (see Chapter Seven).

Photo 5.2 *The procession at an Orthodox Christian festival in Karate town*

Some Konso traders have profited significantly in recent years; a few now own small trucks and grinding mills.

In the mid-1990s, it was evident that the majority of the Konso who were Orthodox Christians were involved in this trade network. The successful Konso traders were particularly conspicuous in their participation in Orthodox Christian festivals. Several also held feasts in their houses on Saints' days, inviting many people from the surrounding community to come and eat meat, in a celebration of, and in thanks for, their health and prosperity.

The Orthodox Church took a much more tolerant position towards local culture than the Protestant Church. The acceptance of Orthodox Christianity did not demand a rejection of the local culture, and many Orthodox Christians still participated in local ceremonies with their *poqallas*. It was not only traders who converted to Orthodox Christianity (although they were the dominant group). It has already been seen how some *poqallas* converted, largely because of the close relationship that they had with the northerners, when they had acted as intermediaries. Membership of the Orthodox religion did not prevent these *poqallas* from carrying out their traditional duties. In some cases, converting to Orthodox Christianity has been used by the *poqallas* to strengthen their traditional role. Kalla, for example, the major *poqallas* who was also an Orthodox Christian, translated *mooha*, the name of the spirit that supposedly visits him at night and advises him, as an 'angel'. In this way, the Orthodox Christian cosmology avoids equating the Konso spirits with the devil (as the

Protestant Christians have done) and instead tries to align itself with the 'good side' of the good–evil duality of the Christian cosmology.

It is clear, therefore, that conversion to Christianity in either form is more than just a spiritual matter; it is also a social and economic one. In both the Protestant and Orthodox cases, joining the church involved joining a network of power, with access to resources, support and assistance. Joining a church also involved subscribing to and promoting a particular discourse with its own practices, ideas, beliefs, and resources. The way in which these networks intersected with the economic and political reforms of the time, and impacted on the agriculture and the reproduction of the landscape, are the main subjects of the next chapter.

Notes

1 The leaders of the Abyssinian Empire were given rank by the Emperor. *Fitawrari* is a politico-military rank which is normally under *Dajjach*, but which according to Bahru implied higher importance than this in the case of Habta Giyorgis. The grading of the titles was from top down: *Negusa Nagast* (King of kings), *Negus* (King), *Ras* (head), *Dajjach* or *Dajazmach*, *Fitawrari*, *Qagnazmach* (literally commander of the right), *Grazmach* (commander of the left), *Balambaras*, (from Bahru 1991: 232–3).
2 Kalzimo Dinote, 1995.
3 Kalzimo Dinote, 1995
4 Rakasa, 9/9/96.
5 Kahano Kalayta and Kahano Kolobo, discussion, 1 December 1995.
6 Kembo, 2002.
7 A Konso man named Iyasu from Jarso is often credited with guiding the Italians to Konso and hence helping them to attack and overthrow the northerners. He is mentioned by Kembo here.
8 Kembo, 9 September 2002.
9 Compare to Merid Habta Mariam in Maale (Donham 1986b: 78–80).
10 Benedikt Jarsonarson, 1996.
11 R.K., September, 2002.
12 A.S., who was comparing experiences of conversion in Haile Selassie's time to his own conversion to Orthodox Christianity in the post-1991 period (9 September 2002).
13 The date given is 1961 Ethiopian Calendar.
14 Estiphanos Berishia, 1996.
15 *Farayta* is a form of control on the fertility of men and women. It could also be interpreted as a mechanism through which households are able to retain control over the labour of younger members of the household. Such features for controlling labour are common in forms of intensive agriculture (Netting, 1993).
16 Estiphanos Berishia, 1996.
17 Estiphanos Berishia, 1996.
18 Estiphanos Berishia, 1996.
19 Estiphanos Berishia, 1996.
20 It seems likely to have varied from one individual missionary to another.
21 Koyiti Kaiya, 1996.
22 When an old man or old woman, who has given birth to a son, dies, then the burial is accompanied by a *shileta* ceremony, in which the person's life is celebrated by dancing and singing.

23 The word for witch, *sawaita*, is better translated as magician, and is used to refer to men or women with special powers and abilities, some of which are negative and harmful; others – like the *sawaita*, who are *poqallas* who can see the future – are generally considered benign.

24 The origins of the term *setana* are obscure at present. It is possible that it is an adaptation of the word 'Satan' but if this is the case, the adaptation may well have occurred elsewhere and some years ago, as the term *setana* is common to several people in the south. Whatever the origins, the similarities between the terms *setana* and Satan are not lost on many Christians who use the similarity as evidence of a connection between the two.

25 Kahano Kalayta and Kahano Kolobo, 14 June 1996.

26 Such an interpretation sees spirit possession and witchcraft as potentially empowering for the powerless and as a form of power that might be the only option to women and other marginalized groups. Histories of spirit possession and witchcraft activities have often associated women and other marginalized people with these activities, which supports this view. It is also possible, however, that witchcraft accusations are made against marginal and powerless groups in order to keep them in a marginal position. The power of witchcraft and spirit possession can thus be used to challenge or to maintain the existing power relations in a community. There is not enough evidence to suggest that women or younger brothers were particularly attracted to *ayana*; its existence was, as is shown later, much more widespread,.

27 'In Borana ... these ecstatic phenomena are so recent that the majority of my middle-aged informants remember when they first appeared in the region' (Knutsson, 1967: 66).

28 Margret Hrobsjartsdottir, 1996.

29 Estiphanos Berishia, 1996.

30 Margret Hrobsjartsdottir, 1996.

31 See Ferguson (1999) for discussion of the importance of 'style', often dismissed as trivial.

Six

Revolutionary State

The 1974 revolution heralded the beginning of a new period in Ethiopia. The *Derg* government penetrated the structures of government more deeply into the grassroots communities of southern Ethiopia than any regime that had gone before. In this chapter, the general history of the revolution is reviewed, setting the context for the discussion of the experience in Konso. As well as exploring the way the revolution impacted on Konso, the way in which the ideas and practices associated with the revolution combined with those of Protestant Christians, described in the last chapter, is examined. The combination of the modernizing forces of Protestant Christianity and scientific socialism had severe impacts on the institution of the *poqalla*. As the modern project failed, however, some Konso began to seek alternative forms of meaning and organization, and returned to some of their older institutions, including the *poqalla*. Here the orthodoxy of customary practices and the institution of *poqalla* show their resilience.

The impact of the revolution and the *Derg* period in Ethiopia

The revolutionary government was known as the *Derg*, from the Amharic word for 'committee'. The events surrounding and following the 1974 revolution in Ethiopia have been described in some detail by Clapham (1988), Dessalegn Rahmato (1984), Ottaway and Ottaway (1978), Pankhurst (1992), Donham (1999) and James *et al.* (2002). The revolution brought some dramatic transformations, including changes in state administrative structures at the local level, and a land reform programme implemented in 1975, which has been described as one of the most radical agricultural reforms that have been set in motion in sub-Saharan Africa (Dessalegn, 1984).

In 1974, Haile Selassie was overthrown in what has been termed the 'creeping coup'. At the time Haile Selassie was over eighty years old, and his mode of government was being seen as increasingly corrupt, outdated and inefficient. As Clapham (1988) explains, he had become a victim of his own success: his overriding aim had been to modernize the country, but this modernization was increasingly at odds with his courtly style of Imperial governance, based largely on the manipulation of personal relations. Modernization had created new demands and expectations, which were largely not met, and new groups of people who were educated and able to formulate and express criticism emerged. The frustrations of the public spilled out in a series of strikes and protests: in Addis Ababa in 1974 'practically every group from prostitutes to lay priests went out on strike' (Ottaway and Ottaway, 1978: 4).

During 1974, the Coordinating Committee of the Armed Forces, Police and Territorial Army formed, with a membership of one hundred and eight military men (Clapham, 1988: 40; Ottaway and Ottaway, 1978: 5). Later known as the *Derg*, it was a secretive group, whose members' identities were protected. It played an increasing role in the orchestration of the revolution.[1] On September 12, 1974, the *Derg* deposed the Emperor, and as Ottaway and Ottaway describe, 'there was no public outcry ... without a shot being fired or a cry raised in his defence' (Ottaway and Ottaway, 1978: 6).

The *Derg* established power as the Provisional Military Administrative Council (PMAC). It appeared to be a revolutionary movement in search of an ideology and coherent set of policies (Ottaway and Ottaway, 1978; Donham, 1999). Their policies had been previously announced by the slogan '*Ethiopia Tikdem*' which is translated by most as 'Ethiopia First', but also as 'Let Ethiopia Move Forward' (Negussay Ayele, 1990). But beyond the slogan, the policies had little content and substance. In December of 1974, they attempted to remedy this by announcing that *Ethiopia Tikdem* means '*hebrettesebawinet*', an Ethiopian socialism which means 'equality; self-reliance; the dignity of labour; the supremacy of the common good; and the indivisibility of Ethiopian unity' (Ottaway and Ottaway, 1978: 63). Their aims in the broadest sense were the elimination of poverty and exploitation (ibid.: 65). As time went on their policies became more radical, and drew more from Marxist-Leninist ideas which fitted well with their own aims and objectives. They included, for example, the need for a strong central authority to lead and liberate the peasantry; it was strongly anti-imperialist and required the eradication of an old order in which the peasantry were 'serfs' who had to be liberated from the oppression of 'feudal landlords'. The Marxist-Leninism that was adopted was therefore another teleological modernist ideology which promised freedom and a better future.

In 1975, banking institutions and many companies were nationalized. Also in 1975, the land reform programme was announced. There had long been pressure in Ethiopia for land reform, which included a campaign whose slogan was '*Meret Larashu*' or 'land to the tiller'. The radical land reform adopted by the *Derg* aimed to undercut the economic power base of the landlords of the old regime. All land was declared the collective property of the Ethiopian people. Ottaway and Ottaway summarize the spirit of the reform: 'The tenant was finally freed from his shackles, and no peasant would ever again have to pay rent or feudal dues' (1978: 11). They also comment that '[t]he proclamation was so radical that it left even diplomats from Communist countries shaking their heads in dismay at the speed with which the military was moving Ethiopia from feudalism to socialism' (ibid.: 8–9).

One difficulty emerged because the new regime abolished the means through which the rural areas of the country had been administered in the past: the rural landlords, or *balabbats*. If these landlords were to be abolished, then who was to implement the reforms? The responsibility was delegated to new Peasant Associations (PAs), which were to be made up of elected representatives of residents from each area of 800 hectares. The responsibilities of the PAs were to administer public property, including land, and to redistribute it where necessary. They were also responsible for coordinating cooperative work projects, which included collective fields, the construction of schools, clinics and roads, and for undertaking villagization programmes, similar to those implemented in Tanzania (Scott, 1998: 223–61; Clapham, 2002; Taddesse Berisso, 2002).

In order to organize the PAs and to teach the 'peasantry' about the aims of the new regime and the virtues of socialism, the government established the *Zemecha* campaign, of 'Development through Co-operation'. This campaign made it mandatory for all 'students', a category that included teachers, university students and those in the last two years of high school, to go to the countryside and take up the revolutionary work. It is estimated that more than 50,000 'students' were sent to 437 locations (Ottaway and Ottaway, 1978: 69). They were to set up the PAs, run literacy programmes, teach about socialism, and run small development projects. This campaign was doubly expedient for the new regime: it filled the administrative vacuum that had resulted from the dismantling of the old order, but it also dispersed from the capital city a group who were increasingly vocal. At the same time, the *Derg* was becoming increasingly ruthless as it crushed what it saw as opposition, either inside or outside its ranks. If the revolution itself had been bloodless, the first few years of the revolutionary government were anything but. The violence culminated in the 'Red Terror' of 1977-8 in which tens of thousands were arrested and many killed. During this time, Lieutenant-Colonel Mengistu Haile Mariam emerged as the Chairman of the *Derg*, and he remained the head of state until the *Derg* was overthrown in 1991.

Revolution in Konso

Kahano Kalayta and Kahano Kolobo described what happened in Konso:

> In 1974 was the next change of government. People came from Addis Ababa, mainly students, but they came to Konso after the change in Addis Ababa, and so by the time they came to Konso the previous northern rulers had already run away, so there was no fighting. One man who came had a gun. The others had sticks. They taught the people, lectured them, and told them how they had fought the previous government and told their new political aims.[2]

Their list of the new political aims of the regime reveals something about how the new regime was understood and experienced.

> [The students said that there would be] no taxes. People can't just come and take your livestock [as the previous regime had done as tribute and latterly as taxation]. You will use your land as you want, and there will be no landlords. The users of the land will be the same as the owners of the land. They set up Peasant Associations so that the administrators were also farmers who knew the problems of the people.[3]

Many Konso people found much that was attractive in the teachings and the promises of the students and the new administrators. First and foremost, they looked forward to a life free from the control and authority of the northern landlords, which they had never fully accepted. They were not surprisingly encouraged by the promises of 'no taxes' and the freedom from the extraction of money and labour which they had long experienced. The quote above shows how the Konso people identified with the anti-imperialist and anti-feudalist rhetoric of the new regime. Clapham may be right in stating that indigenous people in Ethiopia show a healthy scepticism towards political ideologies (1988: 10), but in this case, and at this moment in time, the emancipatory promises of the new regime were close to the dreams and aspirations of local people, and they were therefore taken to heart. In certain ways, the political ideologies were similar to the Protestant Christian teachings: both were promising a future life free from poverty and oppression. Another informant explained about the beginning of the *Derg* period:

> The land was put into the hands of the peasants. They no longer had to pay a landlord in cash or kind. The land they were working they farmed as their own. They were told they had rights and they were mostly happy.[4]

The policies that were implemented following the revolution had economic, political and cultural dimensions. Economically, land was put in the 'hands of the tiller': landholdings of more than ten hectares, share-cropping arrangements and the lending of land to others, were all officially outlawed. Some collectively cultivated fields were established, and each household had to contribute a certain amount of labour to it.

Politically, PAs were established, which were led by a committee of about twelve men, including a chairman, a treasurer, and a secretary. Each PA also had two or three people who are described as 'militia', who carried out the PA committee's orders, and arrested and fined those who did wrong. Officially, country-wide, a PA was set up for each 800 hectare area, and became an administrative unit known in Amharic as *kebelle*. In Konso, almost every village became a *kebelle* with its own PA, regardless of its area. In a few areas, particularly the lowland areas, villagization programmes were implemented. In the majority of highland areas, where people already lived in concentrated settlements, the villages remained as they had been. In the later years of the *Derg*, the PA committee was joined by a youth committee and a women's committee, which took responsibility for youth and women's issues respectively.

The revolutionaries also attacked cultural institutions that they considered harmful obstacles to the establishment of an equal society and their aim to liberate the people. In this way, again, there were similarities between the revolutionaries and the Protestant Christian teachers: both construed 'tradition' as oppressive and as limiting the progress and freedom of the people. In this way, the revolutionaries finished off the work that had already been started by the Protestant Christian proselytizers. In some cases, the customary institutions were also thought to have the capacity for political action, and thus represented a potential threat to the new regime. Attacking and dismantling the cultural institutions thus met several objectives at once.

For example, the generation-grade of *farayta* was attacked as harmful and backward because of its association with abortion. The position of *apa timba*, the institution that enforced law and order in the village, was also attacked. The drums that symbolized this office were in many cases captured and confiscated or destroyed. In some areas, the people managed to hide the drums, and the office continued to function in secret. But in many places, Buso included, hiding the drum meant that it failed to continue circulating, and the *apa timba* failed to continue his work. As the circulation of the drums was halted, and the generation-grade of *farayta* was abolished, the generation-grade system itself became largely obsolete. In most parts of Konso the passing of duties from one generation-grade to another was no longer ritually enacted or celebrated. The generation-grade identity became, as a consequence, almost meaningless, and many of the positions connected to the generation-grade system, for example the *xela*, no longer functioned. In addition, whereas before, the age of first child birth for men and women had often been delayed by the institution of *farayta* of the generation-grade system, age at first child birth reduced as *farayta* was eliminated. The destruction of these cultural institutions has contributed significantly to population growth in Konso, since the 1974 revolution.

The *sawaita*, or, more specifically, those who were thought to be *sawaita*, were arrested as dangerous people who were harming the population. This group of people included *poqallas* with seeing powers, people who practised magic of various kinds, and those with reputations for becoming possessed by spirits. These people were dressed in ridiculous costumes and made to parade around in market places on market days, where people could laugh at them. The revolutionaries also burned the ritual objects and paraphernalia that were used by *sawaita* and those possessed by spirits. By capturing and destroying objects, arresting cultural leaders, and humiliating others, the revolutionaries attempted to smash the authority and power of Konso custom and culture, and replace it with new collectively organized, secular, grassroots social units, which were to provide the basis for a socialist future.

The *poqalla*s embodied many of the aspects of Konso culture that the revolutionaries found most evil. The *poqalla*s were landlords who had more land than others, and who used the control over this land to obtain labour and tribute from others. Many *poqalla*s had also served in the Imperial regime as *balabbat*s and were seen to have benefited from their cooperation. The *poqalla*s were also customary leaders who wielded political power that was seen as both oppressive and as a potential source of opposition. They also carried out rituals and were thus vestiges of oppressive beliefs and superstitions. For all these reasons, attacking the *poqalla*s was seen as a way in which the people could be liberated from the shackles of economic exploitation and the oppression of custom and superstition. An examination of what happened to the *poqalla*s after the revolution reveals a great deal about the nature of the revolution and the impact of the policies that followed. It also explores the fate of the institution which, as I have argued in previous chapters, is central to much of the social organization and production of Konso.

First, several of the *poqalla*s, particularly the regional *poqalla*s, were arrested because of their cooperation with the Imperial regime and their work as *balabbat*s. For example, Bamalle, the regional *poqalla* whose reputation as a harsh *balabbat* has been discussed above, was arrested. His home and granaries were attacked and burned by Konso people who were angry at his actions over the preceding years. He was sent to prison in Arba Minch where he spent 18 months, followed by 18 months in prison in Addis Ababa.[5]

The case of Guufa, the regional *poqalla* for the Fasha region of Konso, shows how the revolutionaries attacked the *poqalla*s because they were cultural institutions, as well as being landlords. In this way, they destroyed their symbolic power as much as their economic power. The revolutionaries set out to desecrate Guufa's sacred status and power, not by arresting him, but by forcing him to break the taboos of his office. Here again, the anti-tradition, anti-culture and anti-superstition elements of the modern revolutionary movement were evident. An old man from Fasha who witnessed the event explained to me what took place:

> From the time of the revolution everything was lost... Guufa lost his land, and also they shaved his hair, and they pulled up his *oluahita* [the structure made from the trunks of juniper trees, where each tree trunk symbolizes a generation that has passed]. They took the stone that he used to sit upon to bless the people and they turned it over. What was left after this that

> they didn't destroy? The old woman [the *poqalla*'s mother] who people even should not touch, they came and took her bracelets from her. They took the bracelet from Guufa [that symbolizes his office]. They cut his hair and they said 'drink!' He should not drink water with others. People who have had any contact with blood should not go near him. Blood should not even fall onto the meat that he would eat [for example during slaughtering]. They made him drink this blood, they made him drink the water. And in this way they ruined him, they destroyed him.[6]

The case of Kalla, also a regional *poqalla*, shows some similarities to the above cases, but also demonstrates that the way in which the revolution developed was not wholly uncontested. As with Bamalle, Kalla's granaries were looted by Konso people immediately following the revolution, but the homestead was not burned. Kalla was not arrested immediately, but he soon came into conflict with the revolutionaries when they came to his homestead and attempted to cut trees from the sacred forest (*mura dawra*) adjoining his house over which he had stewardship. The revolutionaries were especially keen to obtain the valuable juniper trees to use to build their new houses and offices. At this time, wood was also being transported from Konso to Arba Minch to meet the building needs of the new regime there. In Konso, Kalla attempted to defend his forest by shooting at those who had come to cut it. He was subsequently arrested and sent to prison in Gidole, where he remained for one year.[7]

Kalla, however, was a man who had some education and an aptitude for writing letters and for making appeals. While in Gidole, he wrote to Addis Ababa and appealed to the government against his arrest and for the rights to the forest to be returned to him. Cleverly, he appealed to the government's own rhetoric of 'the people', saying that the forest was a place where *all* the ancestors of Konso were buried, and as such it was extremely important for *all* Konso.[8] He also said that his family gave wood to those who needed it, and that it was thus a forest for *all* Konso people. The officials in Addis Ababa answered his appeal favourably and released Kalla. Kalla's forest is the only substantial sacred forest in central Konso to have survived the heavy culling of this period.

Kalla's story also demonstrates that although the revolutionaries attacked various cultural institutions, particularly those that they considered to be threatening or dangerous, their position on 'culture' in general was more complex. Those customs which could be portrayed as 'safe' and important 'for the people' were celebrated, albeit with permission and under the watchful eye of the regime. After all, the *Derg*'s rhetoric did pledge to celebrate the diverse cultures – or nationalities – of their country (Clapham, 2002). One example of such an acceptable and celebrated form of culture was the ritual that accompanied the burial of Kaiyote Kalla, the Kalla *poqalla* who died in 1986. When he died, he was buried in a full public ceremony: his body was embalmed and kept for seven months, a period during which it was said that he was 'unwell'

(Tadesse Wolde, 1992). Here again, however, the perseverance, adaptability and letter writing ability of Kaiyote Kalla's son, Wolde Dawit Kalla, can be seen to have had an influence. Wolde Dawit Kalla himself died in 2003, but earlier he explained to me how he had acquired permission for his father's ceremony, by stressing the importance of the ceremony for the 'people':

> At this time, even some of the administrators respected culture and came to see. Even many foreigners came, and people from Addis Ababa, and saw the culture. They said this is the culture, there is nothing bad in it. We appealed for permission to do it, and said the people love it and want it, and we will do it together with the people. I wrote to the administrators and told them all that it included and they gave permission.[9]

Thus some *poqallas* fared better than others during this period, and this depended on the individual's ability to adapt and present himself and his culture in particular ways to the new regime.

At a village level, where the *poqallas* had a much lower profile, they were not targeted so assiduously by the regime. This reflects the way in which the local *poqallas* had previously played a more minor role in the Imperial regime and had benefited much less from their position as *balabbats* (see Chapter Four). For example in Buso, Kahano Kalayta and Kahano Kolobo explained what happened to the *balabbat*, Aija, following the revolution:

> At this time, the newcomers asked who was the *balabbat* and landlord of the area, and how did he hurt you? They promised to punish him. But the people of Buso protected Aija saying that he had not hurt them, and so he reverted to being just a normal *poqalla*, and also the new regime said that he should live just like an ordinary farmer and that no one should work for him.[10]

Nevertheless, the political and judicial role of the *poqalla* was undermined. The PAs took over the role of conflict resolution and law and order and outlawed other local institutions such as the *xela* and the *apa timba*. Kahano Kalayta explained:

> In the past the clan had great authority over their people, and they even had the right to kill them, and if they did, no one would question the death. By now, even if someone kills one of his clan members, then the government asks, why did the clan kill him? So even the right to live in the village was given by the *xela* and the clan and they administered everything. In the past, before there was a state, the clan had total rights over the lives of their clan members. They can even throw him off a high cliff or beat him.[11]

In general therefore, the cultural and political authority of *poqallas* was undermined. Either their own power as *poqallas* was curtailed or that of the additional offices they held, for example, *apa timba* or *xela* (see Chapter

Four). But one of the areas where the *poqalla* was most heavily hit was in land reform. An exploration of the impact of the land reform, and the shape it took in later years, also demonstrates much about the way Konso attitudes to the *Derg* regime changed over time, as did the impact that the *Derg* had on *poqallas*.

Poqallas and land in the *Derg* years

The land reform proclamation of 1975 made all land in Ethiopia the collective property of the Ethiopian people. All private ownership and the transfer of land by sale, lease, mortgage or inheritance was forbidden (Dessalegn Rahmato, 1984; Pankhurst, 1992; Clapham, 1988). The use of hired labour and landholdings of more than 10 hectares were also outlawed. In Konso, in practice, there was very little redistribution or reallocation of land. This meant that the main impact of the land reform was on those landholdings that had previously been lent to others – the land referred to as *piyolada*, that was lent in return for labour on the fields of the lender, once at sowing, once at weeding and once at harvest (outlined in Chapter Two). It was this *piyolada* which was important in the wider set of exchanges of land and labour between a *poqalla* and others, through which the intensive agriculture and Konso livelihoods were constructed and maintained. After the implementation of land reform, because the previous borrowers continued to farm the same land as they had before, the impact of the land reform could be seen only in the outlawing of *piyolada* labour. *Piyolada* labour became a symbol of the extractory nature of the *poqalla* and became a word that should not even be mentioned.

The *poqallas* no longer had control over large amounts of land, and were no longer able to use it to gain access to the extra-household labour. The weakening of the *poqalla*'s control over land had an impact on their ritual role and symbolic power. The regime made it difficult for *poqallas* to carry out their rituals, because they viewed such practices as retrograde. But the *poqallas*' ritual role was also undermined indirectly by the economic reforms, as the *poqallas* found it difficult to raise the capital to carry out the more expensive rituals.

The economic reforms may have represented a positive development for others who were no longer obliged to respond to the *poqallas*' calls for labour, and who were also encouraged by promises of a time which would be free from poverty and exploitation. As Samuel, quoted above, stated, 'the land [the peasants] were working they farmed as their own. They were told they had rights and they were mostly happy' (Samuel, 21 July 1995).

Careful interviewing and examination of current landholding arrangements showed, however, that this situation did not remain the case for long. After five or ten years, the new state-given landholding rights began to be challenged by those who held customary rights to the land. In other words, the *poqallas* began to re-claim their rights. As Samuel continued:

> After five or ten years, however, the landlords started to get their land back secretly – no one in the administration or the government could know. The landlords would say to people 'I gave you this land because you were in problems, but now I want it back, otherwise I will blame you, you will be offending me.'[12]

A period of much more negotiable land rights ensued, which was made possible because, in practice, there had been no reallocation of land. A person who had originally borrowed land, now had rights to that land because of the state land reform proclamation. But, at the same time, he or she was likely to be in contact with the person who had initially given him or her that land and therefore was susceptible to pressure. This pressure was powerful because the original lending of land was presented, understood and experienced in indigenous ideas as an act of generosity, kindness and assistance. Moreover, if the lender was a *poqalla*, as was often the case, the original lender was likely also to be a relative (the living representative of one's own founding ancestor, no less) or at least a relative by marriage. Acknowledging the original owner's rights to land meant, of course, that the owner had other rights (to labour, for example) in return for that land.

When a *poqalla* said, 'I gave you this land because you were in problems, but now I want it back, otherwise I will blame you, you will be offending me,'[13] it was difficult to resist, for all the reasons discussed in previous chapters. The *poqalla* was central to the spiritual, social and material wellbeing of his lineage. As Kalla put it:

> So after the *Derg*, people said this is my land… but the village didn't like it, [they said] everyone breastfeeds from each other. As you have done this, tomorrow, someone who doesn't have any thing, where will he go and ask for a field to borrow? What will the young men do?[14]

If a person accepted the *poqalla*'s claims to the land, the arrangements that were reasserted with it were not the same as they were before: the labour that was given in return for the land was no longer described as *piyolada*, but was described as *fadeta* (mutual help). In the past, the *poqalla* had only provided *erurota* for the workers on his fields, a cheaper and unfermented grain drink; now the *poqalla* had to supply good strong *chagga* if he wanted the workers to work well. There was also a new flexibility in the labour that was performed on a *poqalla*'s field. It was no longer necessarily carried out systematically three times a year. There was some 'wriggle room': a person who was reluctant to comply with the new claims to labour would be wise not to reject the claims to labour altogether, as to do so would be to offend the *poqalla* and reduce the possibility of applying to the *poqalla* for assistance in the future. Instead, he would merely (and euphemistically) 'forget' to attend, or be engaged in urgent business elsewhere. If a person regularly failed to cooperate, however, it would be interpreted as a statement of refusal. As a consequence, he or she would be asked to return the land to the *poqalla,* and it would be given to someone else.

Although the *poqalla*'s claims to land had been weakened by the state proclamation that all land now belonged to the state, if a *poqalla* asked for land to be returned to him, the request was rarely resisted. I only came across one case in Buso where an individual attempted openly to challenge his *poqalla.* A man who had a *poqalla*'s land tried to sell it. The *poqalla,* with the support of the village, resisted the sale, saying it was his own land, and so the man took the *poqalla* to court accusing him of trying to reassert his claims to land. Taking the case to court was risky, as he himself had tried to sell the land, an act that was also illegal. His court action was unsuccessful and, back in the village, sanctions were enforced on the man. Other villagers refused to give him fire, to work with him, to drink with him, or to herd goats with him. In the end, he gave the land back to the *poqalla,* who then gave it to someone else to use. The man subsequently left the village altogether. It is not, therefore, surprising that land was usually returned.

Buying and selling land was illegal under the land reform and people in the village (men and women) had no compunction in reporting to the government those people whom they considered to be selling land illegitimately. In contrast, land bought and sold was rarely, if ever, reported, when the person selling the land was considered to be the legitimate owner in the eyes of the villagers. Anyone who had borrowed land from a *poqalla,* and who wanted to avoid associated customary obligations, might offer to buy the land from the *poqalla.* Such applications were often accepted, as they were a means to avoid conflict between people. As is discussed later, Protestant Christians, who were loath to have any dealings with *poqallas*, often made use of this option.

From a government perspective, the *Derg* had few powers to tackle unreported land sales or the resurgence of customary landholdings asserted only through the performance of labour. If the administration, for example, found people going to the fields of a *poqalla* to work, they would accuse them of breaking the law, but the farmers would just say 'we are helping our kinsman' or 'friend' or 'neighbour'. There was no way of penetrating this wall of denial, and at the local level at least, the PA committee must have accepted the practice, or more likely, have quietly condoned it.

Disenchantment with the *Derg*

Prior to the *Derg* reforms, the customary network of exchanges that centred around the *poqalla* worked on various levels: spiritual, ideological, social and material. The *Derg* tried to replace this: it tried to destroy the *poqalla*'s ritual power and to replace it with its own ideology; it tried to replace the customary landholding arrangements with land rights given to the tiller; and it tried to replace the customary decision-makers by giving decision-making power to elected representatives on the PA committee. However, it soon ran into difficulties. The Konso people lost faith in its promises, and its new social, political and economic relations on the ground failed to provide the people with viable alternatives.

The problems that accompanied the new regime appeared on many fronts: the collective fields were few in number, but they were resented all the same. Some people contributed more labour, others less, and the yields and profits from the collective fields were rarely seen by those who had laboured. The collective fields came to be seen as just another form of labour extraction in the tradition of the labour tribute demands that had been made by the northern colonizers that had preceded this regime.[15] The PA committees were supposed to be made up of elected representatives, but there were a limited number of people who had the skills to hold the positions of power in these committees. The spirit of electing people who would listen to the needs of their constituency and who could make a difference was soon compromised. In the 1980s, the number of committees at a village level increased, and there were questions about the accountability of some of them. Reports from this time, for example, complain that the youth committee frequently commissioned resources from individuals without explanation or compensation.

The *Derg* regime was also characterized by heavy military campaigns. In the 1970s, the Ogaden War demanded resources and people to fight on the Somali border. In the 1980s, the military activities intensified to new levels in the north of the country, as the government fought against Eritrean and Tigrayan factions. The government scoured the country for young men to send to the battlefields, and conscripted many by force, some of whom were under age. Each Konso village was required to send men.

At first, some Konso men were happy to go to the war front, as they were told they would receive training and payment. The first quota of recruits from Buso was ten men, and volunteers were found easily, some from each neighbourhood. By 1984–5 it became evident that people were not returning from this war, and the PA had to force people to go. Each village had a quota of between five and ten men per year.[16] Many people spent months hiding out in the bush to prevent being sent to the war. Others, when they were taken away, took medicine that made them ill, and they were sent home. It is impossible to know the numbers of people who were lost from Konso during this time, but it is estimated at about 50-60 people per village. Very few of the men or boys who went to the war front returned, and strong resentment against the regime developed. As one old man summarized at a meeting:

> During the time of Haile Selassie and before, they took goats and money to the state, and also we passed our time as slaves. The government of

> Mengistu was the same. It was only different in that they also took children and sent them to the front, and some died there, others returned handicapped, and others, we don't know what happened to them.[17]

In addition to this catastrophe, the taxation demands of the regime increased. The Konso people believed that they had been promised 'no taxes', but the taxation reached new levels year on year. In 1975, a tax was introduced at 4 Ethiopian dollars (*birr*) per household per annum. According to interviewees' memories, increased to 7 *birr* in 1974, 10–20 *birr* in 1977, 60 *birr* in 1984, and was 60–70 *birr* in the 1990s and when the regime fell.

These developments thoroughly discredited the regime and were the context for the reassertion of the customary land rights and re-engagement with the *poqallas*. This also sheds some light on the question posed at the beginning of the previous chapter about how the current practices can be said to be ancient in a situation where the *Derg* revolution and Protestant Christian teachings brought about radical changes. The answer is that indigenous land holding rights and exchanges of labour, in which *poqallas* were deeply implicated, were reasserted (albeit in slightly changed form) and they showed an impressive degree of resilience. The *Derg* regime had failed to fulfil its promises; although the indigenous institutions made few grand claims or promises, they did to some extent function to help support livelihoods.

Nevertheless, not everyone returned to cooperating with the *poqalla*. Some people resisted the reassertion of the customary rights and the recentring of the *poqallas* in the network of exchange of land and labour. Those who refused to acknowledge the rights of the *poqalla* were described as saying '*xayshuma*', which refers to the sorghum stalks that are placed after harvest along the ridges of the square ridged basins on the terraces. For example, Tenayu, an elder from Fasha who lived close to Guufa's compound, explained what happened to Guufa's land: 'The person Guufa gave the land to now says things have changed. He says "*xayshuma*". And so Guufa loses the land.'[18] Those present at this interview explained that saying *xayshuma* is like saying 'because my sorghum stalks are lying on this field, you can see I have been using it in the past, and I will continue to use the land.' In other words, 'I have been investing in its good management and the conservation of its soil and so I have rights to continue to use it.'

Two patterns emerge in an examination of those who resisted the appeal to return to the customary practices, and who said '*xayshuma*'. Firstly, when *poqallas* had become weak, either because they had failed to give birth to children, or they had generally been unsuccessful in their manipulation of the network of exchanges described as breastfeeding, they often failed to reassert their rights. An example of this is Guufa regional *poqalla*, who had almost no descendants and who failed to engender the support of people in his area. *Poqallas* with little wealth and success were not viewed as the source of a viable network of support and there was little positive incentive for the people to cooperate with them. Neither did weak *poqallas* have the resources to put pressure on people to cooperate. It is difficult now to see any difference between such weak *poqallas* and other people in the community. Kahano Kalayta and Kahano Kolobo summarized the situation:

> [These *poqallas*] don't have *piyolada*. They used to give it but the revolution came and the people said this is my land. The people don't respect the family. If the family becomes motherless and fatherless and so decreased, then who are people to respect? A family that is strong, they fight with the people over this. They can fight for what is theirs. The reduced families, they are weak like children, they can't make food, they don't have any grain in the granary. Terrible!
>
> The other families, they are strong and they have plenty of drink, and they make really thick *chagga* (beer) and also food. The people will go to the fields for them today, and tomorrow and even the day after tomorrow. The others don't have anything, they don't give *chagga* and there are no full stomachs. Why should people be concerned for them?[19]

Secondly, there was a group of people who deliberately opted out of cooperation, regardless of the nature and strength of the *poqalla*. This group of people included those who had converted to Protestant Christianity. And it is here that the significance of the changes brought by Protestant Christianity are particularly evident, as is the way they combined with the changes brought by the revolutionary state.

Parallel lives: networks of belief and production

The Protestant Christians, as the last chapter has shown, withdrew from many forms of social, economic, political and ritual life connected with the *poqalla*. All interactions with *poqallas* were seen as tantamount to devil worship, and this extended to the networks of land, labour, and blessings, described by the local metaphor of 'breastfeeding'. In the main, therefore, Protestant Christians resisted the return to the customary arrangements.

The Protestant Christians were able to opt out of the customary networks of obligation for various reasons. Firstly, the land reform did have an impact and weakened the *poqalla*'s claims to land. Although some *poqallas* had managed to reassert their claims, these claims still went counter to statutory law and any claims to land or requests for labour had to be made informally and in a way that was not visible to the state. Protestant Christians, many of whom were either educated themselves, or who had contact with educated people and administrators, were more likely to know their rights and to resist the *poqalla*'s reassertion of claims to land that he had lent. All the same, many Protestant Christians made a payment to the *poqalla* in order to buy the land they were using, and to escape the customary obligations. This payment meant that conflict between the rival claims was avoided.

Secondly, the Protestant Christians had alternative networks of support that they could rely on. Their networks included foreign missionaries, and churches in other Konso villages, in Karate town and further afield. Membership in these networks brought new opportunities, and benefits that were often seen as more valuable and exciting than those available in the indigenous network. They included, for example, education and employment opportunities, and the possibility of attending meetings and rallies in new places. They included forms of educational support, in some cases housing, health care and other forms of assistance.

At the village level also, Protestant Christians did not lose by opting out of the traditional networks of cooperation and labour exchange. Protestant Christians formed their own work groups made up of members of their church, and they worked on each others' fields, or on the fields of others for money and other resources (for example, for wood for building). Some of the money raised through work groups was used to build a small church so that Christians no longer had to worship under the shelter of a tree in the village. Some of the money raised was also used to buy an animal to be slaughtered for the work group to eat in celebration of their hard work, in the same way as the other work groups, such as *parka* and *marpara,* might do. Nowhere is the role of religious affiliation in the organization of the labour more evident than on the day of feasting. Each work group has an animal to slaughter, and the groups are organized by religion because no one will eat an animal that has been slaughtered by a person of another religion. As well as their own religious and ritual frameworks of meaning, Protestant Christians have their own, full, social events and lives.

In the mid-1990s therefore, the changes that impacted on Konso did not lead to a wholesale shift from one form of society to another. They led to the emergence of two main groups of people who were distinguished primarily by their religious identities and beliefs. The religious dimension of the different groups was obvious, and often made explicit, but the membership of each of these groups also entailed different ways of co-operating in agriculture, and different definitions of rights to resources, particularly land. Those who did not convert to Protestant Christianity tended to cooperate with their kin and their *poqalla*, and to accept the land claims of their *poqalla*, despite the 1975 land reform proclamation.[20] These people started to 'breastfeed' again. Those who converted to Protestant Christianity worked together and tended to establish their own independent claims to resources, particularly the land they cultivated. These two groups existed side by side.

Thus, in the mid-1990s in Konso, there were two main ways in which the agriculture and the production were organized. These provided two parallel sets of 'ways of doing things': they included parallel networks of support, social relations, and rights to resources. The defining feature of which network a person followed was his or her religious identity, with clear economic and political consequences. 'Switching' – the movement from one group to another – was possible, but was limited by the sanction that too much switching, or switching without genuine commitment, would be considered insincere. In practice, not everyone would be fully signed up to one group or the other; some may even have wavered between them, but the existence of these two groups, defined in terms of religion, structured the two kinds of behaviour and two forms of organization of landholdings and labour.

These two different 'ways of doing things' and sets of belief can be viewed as rival discourses, competing for legitimacy. Each discourse relates to forms of identity, social networks of cooperation, belief, moral behaviour, and labour and land ownership. Here again, we see how all the different aspects – mundane and sacred – are interconnected. But such an approach also raises the question of the extent to which the term 'discourse' can be used to refer to something as rich, deeply felt and complex as a set of religious beliefs and religious identity. Similarly, does this imply that religious identity and belief is reduced to something that merely bolsters a position of power? Discourse does not necessarily imply insincerity, inauthenticity, or unimportance. Discourses are powerful exactly because they are made up of deeply felt ideas and practices. The discourses may bolster power and dominance, but this does not mean that those who are shaping, producing, and living according to the discourse are motivated only by this.

Religious discourses are extremely powerful as they enable a person to enter a particular world of experience and they also connect people to strong, moral, real and imagined communities. They are powerful because they provide those who practise them with a set of beliefs and practices through which hardship is explained and potentially mediated against. The latter feature of religious belief employs a sociological understanding of the power of religion, and has been described in more detail in Chapter Four. The eschatological aspect of religion provides hope and even promise of a better life to come. This aspect is shared with many other ideologies which are similarly attractive and powerful because of this element. In Ferguson's work on urban Zambia, for example, he describes one group of urban dwellers who believe in the myth of modernity to the point that they expect that they will, sooner or later, benefit from all the commodities and attributes of a modern 'first class' world. They believed that 'with a rising standard of living, bustling urban centres, and such symbols of modern status and suits made in London and a national airline, membership in the "new world society" seemed finally to be at hand' (Ferguson, 1999: 235). The Marxist ideologies of the *Derg* government also promised the imminence of a time in which people would be free from hunger, exploitation and taxes. Both of these promises were betrayed. In Ferguson's case, it was brought about by the collapse of the international copper price. In the Ethiopian case, the *Derg* failed to fulfil the expectations that they had raised. In both cases these dreams were shattered.

The advantage that religious beliefs have over other kinds of ideologies is that when they fail to fulfil their promises in this world, there is always the possibility of the promises being fulfilled in the afterlife. Such is the power of faith, which it is impossible to disprove (Blakely *et al.* 1994).[21] In the mid-1990s, Protestant Christianity was economically and politically strong, and was going some way to providing a better life for many in the present. Those who continued to experience hardships could at least hope for another less painful life in the future. The indigenous beliefs and practices, surrounding the *poqallas*, were not so strong, but they promised much less; and thus their followers were not liable to suffer from disenchantment. They continued to provide a means of mediating against hardship in terms of blessings, rituals, economic and political support. For such reasons, these religious discourses and networks have endured, and have continued to shape the organization of agriculture in Konso.

In recent years, scholars have shied away from examining links between religion and production, because such arguments are associated with functionalist approaches which would stress that religion serves a purpose in relation to obtaining a living in a particular environment, and that it holds society together harmoniously. Such approaches have been heavily criticized for focusing on a holistic and bounded community, for treating religion in terms of the purposes that it serves rather than the meanings that it holds for people, and for presenting a timeless, unchanging picture. Criticisms of this approach are entirely valid, but the critique must not be taken to the point such that studies fail to recognize the power of religious ideas and practices, and their connection to other aspects of life. Religious communities and religious beliefs have a powerful role in structuring and galvanizing action, and conversion to a different religion can involve a change in a wide range of relations and practices which are not only spiritual, but are also practical and economic. A study of religion should not concentrate exclusively on its repercussions, but neither should it ignore these more practical, political, and economic dimensions. To study it from the perspective of a discourse enables an appreciation of the way in which religion is deeply powerful and interweaves different forms of speech, style, belief, morals, institutions and practices. It also allows an exploration of the way in which these different practices come into dominance over time, as the result of a combination of endogenous and exogenous processes of historical change and intersection with other discourses.

Conclusion

This chapter, and the one that preceded it, have traced the impact of two major modernizing forces in Konso: Protestant Christianity and the state. They have shown how the two combined to have a profound impact on the society and the organization of production. They have shown that the 'life-cycle' of the *Derg* ideologies and policies was relatively short. People were initially convinced by their promises, but as they failed to be realized, the people become disenchanted and distrustful of the state. One government official summarized the situation in the mid-1990s:

> Even now they [people] suspect the government. They think they [government] are exploiters, [they say] 'They will deprive us of our rights, they will take us out of our culture and they will not lead us in a good way.' Always they suspect because they have been through many cheating governments, you see. They say government is just a means of cheating people because they have experienced it form Menelik to *Derg*, and so they are suspicious.[22]

As the modernizing promises of the *Derg* failed, many returned to the customary networks of ritual, land management and labour organization, which were outlawed during the *Derg* years. Although these had their problems, they were socially and symbolically meaningful, and also provided much needed forms of support.

The *Derg* policies did have an impact, however, in that they provided an alternative form of land management and labour organization that was taken up by another major movement, the Protestant Christians. The land and labour management of people in this group did not follow the *Derg* (and post-*Derg*) policies strictly. They did not, for example, reject an individual's claims to own the land, or to buy and sell it him or herself. But it is unlikely that these more individualistic landholding arrangements could have developed and been put into practice by the Protestant Christians if the *Derg* policies had not undermined the *poqalla*'s claims to land by outlawing it during their rule.

Change has taken place in Konso as Konso people have taken up the opportunities presented to them by the changing national and international context, but the Konso people have appropriated and transformed the different elements of these opportunities and made them their own. This process has not resulted in a breakdown of the institutional dimensions of the indigenous intensive agriculture, but it has resulted in two main forms of organization of land and labour management. These two forms have been differentiated mainly by their respective practitioners' allegiance to particular religious ideas, beliefs and institutions. The Protestant Christian religion, led by *Mekana Yesus*, has proved to be particularly potent and attractive. The religious beliefs in which the *poqallas* have played a central role, in contrast to Protestant Christianity, have not been so coherently expressed and advocated. But these indigenous beliefs, and the practices that they are connected to, have proved resilient and adaptive, and have endured. In contrast, ideologies without this dimension of faith, such as that of the *Derg*, have proved much less powerful and more short-lived.

In her work, Berry has shown that 'membership and status in various social units' determine access to land, labour and capital needed for agricultural production (Berry, 1989b, 41). The membership of social units in Konso has also been shown to be important in gaining access to these resources, and shaping the use of them. In this work, it is the social units and networks structured by religious allegiance and identity, mediated by morally weighted ideas about what is modern and what is traditional, that have been shown to be critical to the way in which the land and labour are managed.

What is particularly interesting about this is that the profound changes that have taken place over the last century have not led to a breakdown of the institutional dimensions of the indigenous agriculture, as is often feared. The indigenous forms of organizing land and labour were undermined severely, but they have been reinstated in a changed form. New forms of organization of land and labour that have local support and legitimacy have also been created.

There are, therefore, multiple forms of organization and land rights which are tolerated on the ground. The potential for conflict between these two different ways of doing things would appear to be high, but it has already been seen how the two groups have affected a compromise by allowing Christians to 'buy out' of their traditional networks and obligations. Conflict and religious differences frequently go hand in hand, but in this case at least, it appears that allowing economic and political rights and claims to be expressed in terms of different religious identities has created a situation in which the different claims and methods can be accepted. The assertion of different rights is not seen as something that is carried out by an individual for his or her own personal gain, but because of his personal commitment to a set of powerful religious ideas and beliefs. Another person may not agree with these beliefs, or may not like the consequences, but they are accepted as a matter of personal choice.

It is possible that the existence of multiple claims to land and multiple forms of organization is also tolerated on the ground for other reasons. At present, it is 'received wisdom' among policy-makers that tenure security is a key determinant of good agricultural practice, environmental management and efficiency. According to this view, seemingly chaotic arrangements in which there are multiple claims to land are problematic. They are thought to result in insecurity, environmental degradation, landlessness and increasingly costly (in time and money) cases of litigation. This may be true. For the Konso, however, the multiple claims to land and the multiple networks of labour cooperation may also be advantageous for some. While pressure on land is increasing and land is becoming increasingly scarce, the existence of different networks provides more potential sources of assistance which, for the poorest in particular, could be advantageous. In addition, as compromises over landholdings between groups are made, multiple claims do not necessarily result in tenure insecurity.

Notes

1 The disintegration of the Imperial regime that took place that year is symbolized by a film shown on state television in September of that year. The film contained footage recorded by the British journalist Jonathan Dimbleby of the famine that had taken place in the north of the country in 1972. The pictures of famine victims were counterposed with inserted 'scenes showing the Emperor and his entourage drinking champagne, eating caviar, and feeding meat to his dogs from a silver tray' (Ottaway and Ottaway, 1978: 57). The showing of this film reveals the extent to which the Emperor himself had become discredited, and the level and organization of the opposition against him.
2 Kahano Kalayta and Kahano Kolobo, 1995.
3 Kahano Kalayta and Kahano Kolobo, 1995.
4 Samuel, Dokato, 1995.
5 This is one of the few cases in which the participation of Konso people in revolutionary acts was discussed openly. This may be a reflection of a lack of participation of people more widely, but also of a lack of willingness to admit culpability for the revolutionary actions. The latter is particularly likely as, when the interviews for this research were carried out, the rhetoric of the *Derg* regime had been discredited. In the majority of other cases, including the ones that are discussed presently, the actors were generally discussed as 'students' or 'soldiers'. Those who led the administration at this time in the headquarters at Bekawle were largely people from outside of Konso. Any Konso people who were educated and who could serve the new regime in any way were sent to work in other areas.
6 Kaitari Tenayu, 1996.
7 The man who was arrested was Wolde Dawit Kalla, whose father was an old man and was still *poqalla* at that time. Wolde Dawit became *poqalla* after his father's death in 1986. As Wolde Dawit's imprisonment coincided with a period when he was not officially *poqalla*, his imprisonment did not mean that he broke his taboos of office, or that he was symbolically spoilt by the experience.
8 Kalla's ancestors are buried in this forest. It is technically only the place of all Konso people if it is accepted that Kalla is descended from the founding ancestor of all Konso people (see Chapter Two).
9 Kalla, 1 August 1996.
10 Kahano Kalayta and Kahano Kolobo, December 1995.
11 Kahano Kalayta and Kahano Kolobo, 1 June 1996.
12 Samuel, Dokato, 1995.
13 These words are Kalla's, from an interview in 1996.
14 Kalla, 1996.
15 Konso people did not mention fixed price or compulsory grain purchases, a much disliked feature of *Derg* policies elsewhere.
16 If members of the PA were ever convicted of bias or taking bribes in choosing the men, they had to send a member of their own family.
17 Elder from Olanta village, at political meeting, 5 December 1995.
18 Tenayu, 1996.
19 Kahano Kalayta and Kahano Kolobo, 14 June 1996.
20 Some *poqallas* also converted, but it is notable that they often took up quite senior positions in the church, thus possibly exchanging one kind of symbolic capital for another.
21 Blakely *et al.* (1994) note that this 'unfalsifiable' element is one of the defining features of religion.
22 KG, 1996.

Seven

Ethnic Decentralization & Self-determination

The previous two chapters explored the way in which the values and practices of the *Derg* government and Protestant Christians intersected. The relationship between these two bodies was not straightforward as they did not share the same values, but they shared some of the same goals. They shared a desire to eradicate aspects of culture they found harmful, and they were both modernizing in their own ways. The outcome of this convergence of aims was a 'discourse coalition' (Hajer, 1995). The Protestant Christians became some of the main supporters of reforms brought in by the *Derg*. As a result Konso society became bifurcated. Two groups of people, discourses, and 'ways of doing things' emerged, which are described here as corresponding to Bourdieu's orthodoxy and heterodoxy. The orthodoxy was made up of customary practices based in the belief that 'they are the way things have always been'. They had the legitimacy of 'tradition', the power of the belief in '*das "ewig Gestrigen"*', 'the eternal yesterday' (Weber, 1978). The heterodoxy was made up of a combination of appropriated *Derg* policies and Protestant Christian teachings. Whereas the orthodoxy involved a commitment to the indigenous institutions of the *poqalla*, to customary tenure and to indigenous networks of labour access, the heterodoxy involved a commitment to state-defined land tenure, and rejected indigenous practices as 'devil worship'. The heterodoxy saw itself as 'modern', the customary as 'traditional'. Modernity (as well as the idea of 'tradition' which is opposed to a 'more advanced' 'modern' condition) had indeed become a model that was constructed and held locally. The substance of the categories 'modern' and 'traditional' were defined *in situ*, and encompassed strongly moralistic ideas of good and evil derived from appropriated forms of Protestant Christianity. The values embodied in each of the categories were used to give value and meaning to other practices.

Following the 1991 change of government, ideas, values, attitudes, practices and arrangements changed again. They did not change immediately, as the new context and new government policies took time to filter through to Konso. When they did arrive, the changes re-drew the alliances between groups, and restructured the constellations of power. New actors came on to the scene, complicating the picture. The certainty and strength of the *Derg* government had already waned before 1991, but, in the early post-*Derg* years, the Protestant Christian movement was still strong and dominant. As the 1990s passed, the certainty and power of the Protestant Christians also began to ebb. The new government was, at the same time, more in favour of custom and 'tradition', issuing statements that called explicitly for indigenous culture to be respected.

These new developments disturbed the older patterns: the simple dichotomies that had developed between 'modern' and 'traditional', and between Protestant Christians and the culture of the *poqallas*, were shaken up. In the place of these dichotomies there were multiple ways of doing things, many new ideas, and multiple actors and possibilities. There were new alliances between previously opposed groups and discourses. In this new, more complex situation, models like that of Bourdieu, with its opposed heterodoxy and orthodoxy, no longer hold. In order to capture this new complexity, as the last two chapters have been described as the modernization of Konso and its agriculture, this chapter can be described as charting their postmodernization.

If caution must be exercised in applying the terms 'modern' or 'modernization' to the Konso experience then it may seem foolhardy to apply to it the term 'postmodern' or 'postmodernization'. The 'postmodern condition' is associated with a highly specific historical experience, which would appear more relevant to life in urban Los Angeles (Soja, 1989) or to studies of the cinema (Harvey, 1989) than to life in agricultural Ethiopia. Its association in my own mind with Konso began simply as a way of indicating a period of time and set of processes that developed *after* those 'modern' processes described in the last two chapters. Although the term 'postmodernization' emerged as a label of convenience, it is also appropriate for other reasons. As well as the disruption to older cultural patterns, this period has been characterized by more diverse and fragmented influences and discourses. There is less clarity to the different world views. Hence, just as the period in Konso discussed in the last two chapters shared many features with other places and experiences that could be described as 'modern', the later period that this chapter describes shares many features with other places and experiences that have been described as 'postmodern'.

The definition of 'postmodern' and the delineation of its features is an elusive task, however, and some authors have gone as far to argue that the postmodern defies a unitary definition by its very nature (Benko, 1997). Others have focused on whether or not the 'postmodern' represents a radical shift from the 'modern' economic, cultural and political mode of life, and have argued that it represents a stage in which traits already evident in modern lives (revolution, invention, uncertainty) have merely become faster and more extreme (Berman, 1982; Harvey, 1989; Jameson, 1991). Eagleton describes postmodernity – that has emerged in a specific postmodern historical moment – as

> a style of thought … [which] sees the world as contingent, ungrounded, diverse, unstable, indeterminate, a set of disunified cultures or interpretations which breed a degree of scepticism about the objectivity of truth, history and norms, the givenness of natures and the coherence of identities (Eagleton, 1996, vii).

There is no need to enter into these debates here. But it is possible to highlight some features which are evident in the quotation above, and which are relevant to the Konso experience.

One feature concerns the demise in the belief in certainty and in single, unifying narratives or theories. The usual example given is the rejection of an unconditional acceptance of scientific ideas and theories. But a similar scepticism is evident towards any grand, totalizing theories or narratives (Benko, 1997). In the context of Konso, this could apply to the way in which Bourdieu's model is no longer relevant, nor is the valorized opposition between 'modern' and 'traditional'. In Konso, a scepticism has also become particularly evident in relation to promises of a better life, or the claims to have the solution to the problems that are experienced. The Konso people have lost faith in the promises of the *Derg* and Protestant Christians. This is also the sense in which Ferguson uses the idea of postmodernity as a time of disillusionment, crisis, or breakdown. The Zambians he describes, for example, were experiencing a kind of post-modern crisis and feelings of abjection as their 'expectations of modernity' had failed to materialize: 'they have increasingly become postmodern; a cynical scepticism has replaced an earnest faith when it comes to the idea of a modernizing, progressing Zambia' (Ferguson, 1999: 14). In Konso, a certain scepticism has also arisen towards the promises of salvation of earlier years.

A second and related feature is the increase in diversity and an emphasis on relativism and difference (Benko, 1997). This relates to the multiplication of ways of living, forms of expression and organization. It also relates to the increase in the number of actors, and to new forms of struggle against mainstream or dominant forms of organization. A third feature is a further emphasis on identity and on identity politics. This emerged as important in the previous chapter, as 'who you were' and 'what you believed in' influenced the way in which land was managed and labour performed. But the identity politics are now more self-conscious, and there are, again, more options and possibilities. The definitions, roles and status of particular identity groups have also been challenged, as the boundaries between identity groups are blurred.

The labelling of this section as the 'postmodernization of intensive agriculture' started, therefore, as a device to indicate another set of processes of change from the one described in the last section. This label can remain as a mere device, as the chapters in this section outline the changes in forms of organization, in attitudes, values and beliefs, and in livelihoods, that have occurred since 1991. But the nature of the experiences in Konso that are described here has a very different quality from those that have been described in previous chapters. In this chapter, the changes brought by the post-1991 government are first outlined. Then, the impact that these changes have had on Konso in general, and on the *poqallas* in particular, are analysed.

To some extent, the post-1991 period in Konso fits what Chabal and Daloz (1999) have described as a 'retraditionalization' of society. Over this period, many indigenous institutions, including the *poqallas* (and the other positions they often hold, like *apa timba*) have been reinstated, and in some cases they have been literally unearthed. The factors that have been important in this process have included the climate created by the post-1991 government, in which all cultures can be celebrated and in which everyone has the right to live as he or she wishes. Also important has been the failure of the modern discourses of the state and of Protestant Christianity to fulfil their promises. These developments have meant that people have returned further to their customary networks and practices.

The revival of the *poqallas* has been limited by certain factors, however. Most significantly, the retraditionalization process has not been neutral. The *poqallas* always were institutions of power and this has meant that reviving their role has had wider ramifications. In the current climate of multi-party elections, the institutions have become politicized. In addition, wider changes to power relations in Konso that have taken place at this time have enabled a certain degree of resistance to the full revival of these customary institutions.

The 1991 Revolution

By 1990, the *Derg* government was weak on many fronts. It had failed to deliver on its promises, and many people, from the military officers to the bureaucrats, to the population at large, had become disillusioned with the regime (Clapham, 1992; Donham, 1999). The government had also been suffering defeats in its war against insurgency groups in the north, as the two main protagonists, the Tigrayan People's Liberation Front (TPLF) and the Eritrean People's Liberation Front (EPLF), had joined forces and were enjoying military success. In May 1991, Mengistu Haile Mariam, the leader of the *Derg*, fled into exile in Zimbabwe, and the Ethiopian People's Revolutionary Democratic Front (EPRDF), a coalition of opposition groups, took control of the country. The EPRDF was dominated by the TPLF, and the new EPRDF leader was also the prominent TPLF leader, Meles Zenawi. The EPRDF established a transitional government, which was to run for two years, during which time elections would take place and a new constitution would be ratified. The ratification of the new constitution and the formal inauguration of the Federal Democratic Republic of Ethiopia (FDRE) took place in 1995 (Fullerton Joireman, 1997). Until the time of writing, the EPRDF has remained the ruling party, and it continues to be led by Meles Zenawi, who is Prime Minister, and also it continues to be dominated by the TPLF (Young, 1996; Clapham, 1992).

The EPRDF government, or *'ehadeq'*, heralded a new era in various ways. First of all, it gave a primary role to ethnicity – or 'nationality' as it is referred to – in the structure and management of the Ethiopian state. It instigated an ambitious and radical programme of decentralization and created a new federal state organized along ethnic lines. Its approach is best summarized by Article 39 of the new Constitution which proclaimed that 'every nation, nationality and people in Ethiopia has an unconditional right to self-determination, including the right to secession' (FDRE, 1995: 13). Each 'nation, nationality or people' was defined as 'a group of people who have or share a large measure of a common culture or similar customs, mutual intelligibility of language, belief in a common or related identities, a common psychological make-up, and who inhabit an identifiable, predominantly contiguous territory' (ibid.: 14; Article, 39: 5). Each nation, nationality or people was to have the right to define its own rules and regulations, and to select and implement its own policies. The 'right to secession' signifies the serious extent of the decentralization process, although the EPRDF hoped that it would prove more attractive for ethnic groups to remain as part of the Ethiopian state than to separate from it.[1] According to the decentralization process, each ethnic group was made responsible for the administration of its people, including its judicial system, and given responsibility for all government matters outside those of defence and foreign policy (Fullerton Joireman, 1997). The Ethiopian state was restructured into twelve regions, each representing one large ethnic group, or, where no large ethnic group existed, a collection of smaller ethnic groups. Each region had its own regional government and, over time, developed its own constitution.

With this policy, the EPRDF was carrying out an ambitious plan that resembled to some degree the 'ethnic state' advocated by Chabal and Daloz (1999). For Chabal and Daloz, an 'ethnic state', which centralizes ethnicity in its structure and organization, represents an exciting possibility because it allows the state to build on what are seen as the positive moral dimensions of an indigenous form of social organization. Building on Lonsdale's concept of 'moral ethnicity' (Lonsdale, 1996), they argue that ethnicity could provide the basis for a modern state which is both African and accountable to its people (Chabal and Daloz, 1999: 56–62). By centralizing ethnicity in the state, and by decentralizing a degree of power to each ethnicity, the competition between different ethnic groups for political power and resources could, in theory, be avoided. This competition is frequently associated with 'political tribalism', the ugly side of ethnic politics which has been problematic in many other African

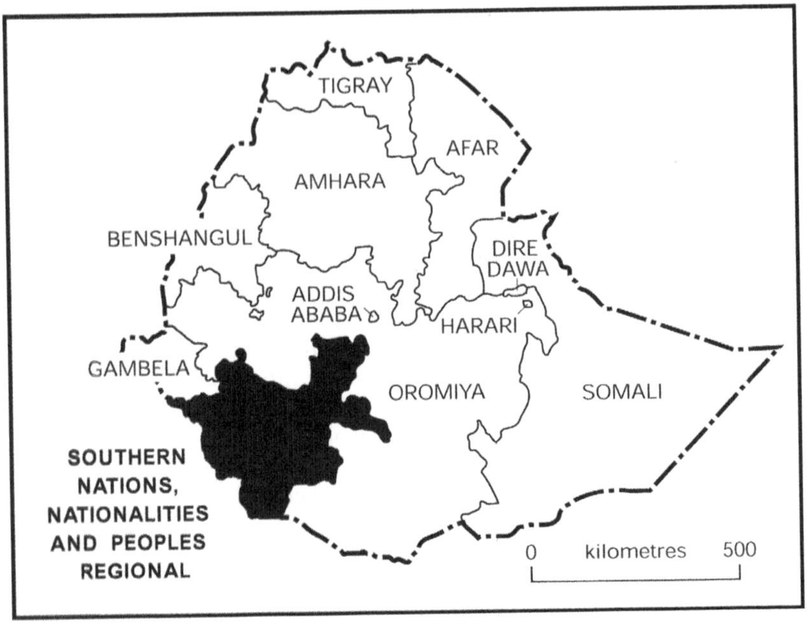

Map 7.1 *Regions of Ethiopia and the location of the Southern Nations, Nationalities and Peoples Regional (SNNPR) State*

countries (Lonsdale, 1996; Chabal and Daloz, 1999). Whether or not this thinking influenced the EPRDF directly, the experience of this process has shown that the 'ethnic state' is not necessarily easy to institute; it does not automatically enable some innate moral ethnicity to be incorporated usefully into state structures; nor does it eliminate competition between ethnic groups (Pausewang *et al.* 2002; James *et al.* 2002; Turton, 2006).

The EPRDF were more influenced in their adoption of these policies by their own domestic experience of the TPLF as a successful resistance movement. The TPLF was based in Tigray, a relatively ethnically homogeneous region that had been effectively cut off from the rest of the country during the sixteen-year war that it fought against the government (Clapham, 1992; Young, 1996). During this time, the Tigrayan people had organized their own forms of local government and managed education, health care, food distributions, and military operations. They had developed a strong sense of self-reliance and pride, and a highly organized and effective degree of grassroots participation in their system of local government. Young summarized the general view that was held by quoting a 'peasant' who told him it was felt that 'only Tigrayans could solve Tigrayan problems' (ibid.: 353).

The TPLF-dominated EPRDF was also reacting against the years of highly centralized rule under the Imperial government and under the *Derg.*

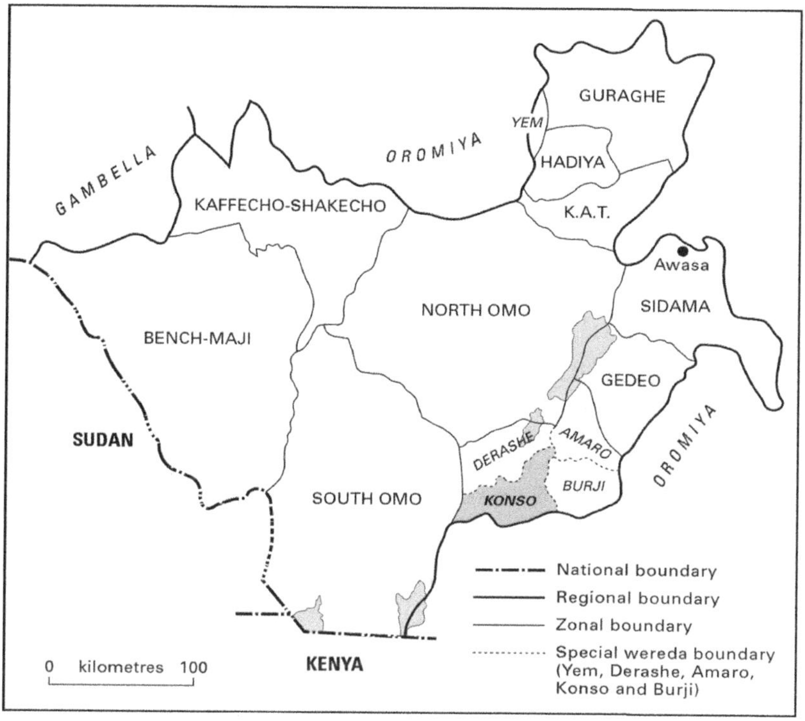

Map 7.2 *Approximate location of Konso special* wereda *in SNNPR State*

Under these regimes, ethnicity was subsumed under the broader project of nation building, but, at the same time, certain ethnic groups were assigned more privileged status than others. More specifically, Amhara culture, religion, and language were favoured, as the religion was equated with the religion of the country more generally, and the language was afforded official status. By contrast, the languages, cultures and beliefs of other ethnic groups in the country were subordinated and presented as inferior (Krylow, 1994). Even though, in theory, the *Derg* recognized the equality of all the different ethnic groups, cultures, and languages in the country, in practice the superior status of those of the Amhara remained (Young, 1996).

Inevitably, setting up this new state has not been an easy process. Since the first restructuring of the state, the boundaries and delineation of these regions have changed. The new Federal Democratic Republic of Ethiopia (FDRE) is now organized into nine regions (see Map 7.1). In addition, two of the largest cities, Addis Ababa and Dire Dawa, have their own charters and a certain degree of autonomy in terms of policy. The redrawing of the

Ethiopian map along ethnic lines, and the establishment of these decentralized ethnic units of administration, has been most straightforward where there are large, homogeneous, and geographically concentrated populations of one ethnic group (for example, in what are now Tigray State and Amhara State). The regional state in which Konso now sits, the Southern Nations and Nationalities Peoples Region, is more complicated: it is made up of many different ethnic groups, and its headquarters are in Awasa (see Map 7.2). In this case, a balance has to be made between the organizational and administrative processes at a regional level, and the rights of each smaller group like that of Konso, which qualifies as a 'nationality' but which exists at a sub-regional level, to govern itself.

The decentralization policy has had profound implications, both politically and culturally. Politically, it has given groups, previously marginalized from the structures of power, a greater say in their own governance. It has been heavily influenced not only by the feelings that emerged from the Tigray experience, but also by international discourses which have been promoting grass-roots participation, decentralization and the rolling back of the state. Culturally, these policies have also aimed to create a context in which what the Constitution calls the 'rich and proud cultural legacies' of the country can be celebrated (FDRE, 1995: 2). Now 'Every nation, nationality and people in Ethiopia has the right to speak, to write and to develop its own language; to express, to develop and to promote its culture; and to preserve its history' (FDRE, 1995: 13). This celebration of local cultures aims to reverse the years in which these cultures were considered inferior. Moreover, culture is seen as something that can be encouraged, or even harnessed, for the development of a new and diverse multi-cultural, multi-ethnic Ethiopian state.

In addition to the emphasis on decentralization of power and ethnicity, the new state has placed a strong emphasis on democracy and equality. There has been a great emphasis on political freedom and the right to vote. This culminated in mass participation in multiparty democratic elections in 2005, which were accompanied by great levels of optimism from the populace. Unfortunately, the government contested the first election results in which many opposition candidates won their seats. As people protested against the government's rejection of the results, many thousands were arrested, some killed. Violence, deaths and arrests have also accompanied the electoral process in Konso. The state's emphasis on political freedoms has been shown to be mainly rhetorical, and political crackdowns have been more the norm, across the country and in Konso.

The post-1991 government has also gone some way to adopting neoliberal market reforms. The expectation is that, by deregulating markets, they will generate a more conducive environment for private business enterprise and for foreign investment. The government signed up to a World Bank 'Poverty Reduction Strategy Paper', known as the 'Sustainable Development and Poverty Reduction Strategy Paper', which aims to enable Ethiopia to meet the Millennium Development Goals[2] through a combination of growth in agricultural and industrial sectors. These papers recommended reforms in the commercial, budgetary and judicial sectors of the country. However, up to now the government has resisted calls to reform the land markets, and to privatize land. It has made some concessions: mortgaging, leasing and sharecropping arrangements of land are now permitted. With these significant exceptions, the situation with regard to land remains similar to the way it was under the *Derg*. The Constitution states, for example, that 'The right to ownership of rural and urban land, as well as of all natural resources, is exclusively vested in the State and in the peoples of Ethiopia. Land is a common property of the Nations, Nationalities and Peoples of Ethiopia and shall not be subject to sale or to other means of exchange' (FDRE, 1995: Article 40, 14).

The policies outlined above show that the EPRDF espoused, in rhetoric at least, a commitment to the dominant values of the international political order of the 1990s and 2000s. They emphasized democracy and participation. They also emphasized the rights of people from each ethnic group to live as they wished, and in this way, like other postcolonial peoples, they were struggling to find ways of living and defining themselves which were different from those imposed by dominant colonizing cultures (Escobar, 1995; Rahnema and Bawtree, 1997). Economically, they also went some way to meeting the neo-liberal agenda by deregulating markets (Peet and Watts, 1993). The history of the regime so far suggests that, on the ground, the results of these policy shifts have been mixed.

Decentralization in Konso

These developments have had profound implications for the people living in Konso. The Konso population was included with other small ethnic groups in the Southern Nations, Nationalities and Peoples Regional State (SNNPR) (see Maps 7.1 and 7.2). Below the administrative level of the region, there were two more levels of government (not including the Peasant Associations that continued to function in some form at the grassroots village level). The first was the intermediate level of the zone, and below this, the smaller unit of the *wereda*. In the SNNPR, some ethnic groups were considered large enough to have an intermediate level of government at the zonal level. Others, where the ethnic groups in question were small, were grouped together with other ethnic groups with whom, it was argued, they shared linguistic and cultural characteristics, and administered together in a zone (for example, South Omo). Konso, however, was considered too ethnically distinct to be grouped together with any of its neighbours, and too small to be a zone, and hence, along with half a dozen other groups of people in the country, it was given the status of '*liyyu*' or 'special' *wereda* (Clapham, 2002; Watson, 2002). Each

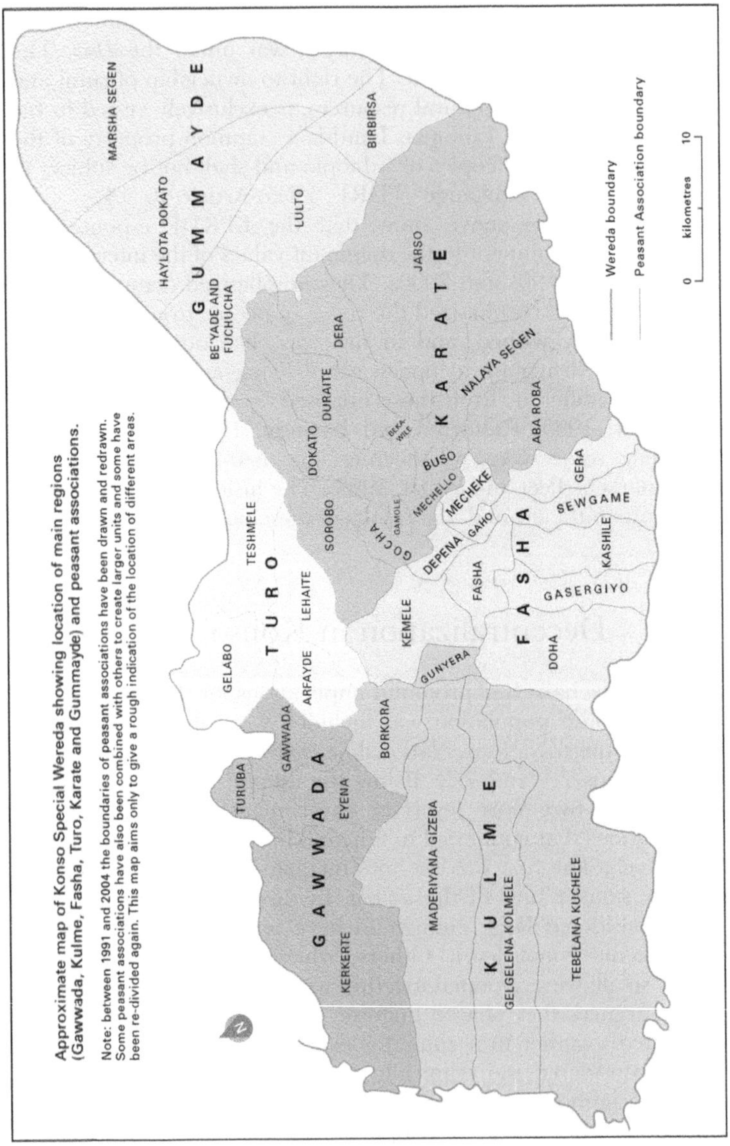

Map 7.3 *Sketch map of Konso special* wereda

special *wereda* in the SNNPR had the administrative powers of a zone and deals directly with the regional administration in Awasa. As a nationality of its own, it has the power to make its own rules and regulations, its own policies, and to distribute its own budget. As an official in the new Konso government explained to me in 1996:

> We work mostly with Awasa but according to the formation of the state, the power is given down, the power is more in the bottom than with the state. More power is given to Konso, even the people themselves have the power to decide.[3]

For people who had been previously excluded from decision-making structures and subjugated by the state, this was a time of great hope and excitement (Watson, 2002). In this aspect, this revolution was not unlike the 1974 revolution that had brought the hopes and expectations of 'no taxes' and freedom from oppression; but at that time the Konso people continued to be ruled by outsiders. Now the Konso people were to be able to rule themselves, and, for some, this was understood as bringing the possibility of a return to a 'golden age' in Konso, to a time before Konso was captured and ruled by outsiders.[4]

The most immediate evidence of the changes that the status of special *wereda* brought was in the employment opportunities that it made available. All the offices of health, agriculture, education, finance, culture, the Konso council (which headed the special *wereda*), and the judicial courts, needed staffing. The official language of these offices was now *Afa Xonso*, the Konso language, for which, at the time of the establishment of the special *wereda*, no formal system for writing had been decided. Despite this, people who were fluent in this language were required to fill the positions in government offices and in schools.[5] Under previous regimes, the few educated people from Konso had been sent to other regions; now these people came back and took up positions in local government. The use of the local language in the offices had practical and symbolic significance. Practically, it meant that any Konso person could come and ask a question, request assistance, or make a case. The people in the Konso villages agreed that this was a profound improvement as they no longer had to 'buy translation' when they had to go to the courts, or to the police, or to the office of the Ministry of Agriculture. The people in the villages also saw that those who were being employed in these offices were Konso people, and they were no longer afraid to approach them. They referred to the new government officials as their own 'children', as they had been born in Konso. The changes were also symbolically important, as they raised the language and the local culture to a status equal to that of other official languages and cultural practices in the country. In fact, it could be argued that, as a 'special' *wereda*, its culture and language was given a higher status and greater promotion than in areas where the culture and language of minority groups had to compete with others in a zone (Watson, 2002).

In many ways, these processes were understood and experienced as a reversal in policies on the part of the government. They had wide implications for the institutions, patterns and processes that have already been described in this book. They opened up space and opportunities for customary institutions such as the *poqallas* to play a stronger and more public role again, particularly as changes also took place in the other force that had been responsible for undermining *poqallas* in previous years, Protestant Chrstianity, as practised in Konso.

Protestant Christianity in the post-1991 period

Under the *Derg*, the Protestant Christians of the *Mekana Yesus* church were powerful, but they worshipped and managed their church under certain constraints (Eide, 2000). In Konso, church services could only be held in Karate (then Bekawle) in the church in the mission compound, and church leaders suffered from intimidation, persecution, and, at times, imprisonment. This pressure limited the level and success of their proselytizing activities, but it did not, it appears, weaken the resolve of its converts. As discussed above, there were useful points of alliance between the agendas of the Marxist revolutionaries and the Protestant Christians. In addition, although the practices of the Christians were restricted under the *Derg*, evidence from other parts of the world has shown how the conditions of adversity can, paradoxically, strengthen the determination of the followers of a religion.

After 1991, the new government's emphasis on democracy and decentralization included an emphasis on the rights to freedom of expression, to freedom of association, and to freedom of religion.[6] The *Mekana Yesus* leaders, who had worked for a long time under difficult conditions, were well placed to take advantage of this change in circumstances. A team of evangelizers was employed by the church and went by foot or motorbike to even the remotest villages, where they talked to people, taught them about the Bible, and gave sermons. In the early years following 1991, *Mekana Yesus* Protestant Christianity had fresh impetus and the movement was at its height. In the villages, older converts were happy that they could now practise their religion freely. There were also many new converts, many of whom were young people.

Soon, however, serious problems arose in the ranks of the *Mekana Yesus* church. The problems were ostensibly about styles of worship and religious principles. They became manifest in a dispute between older, more established leaders of the church, and new, younger leaders. The older leaders favoured more conservative forms of worship, particularly where styles of music were concerned. The two groups inevitably became embroiled in bitter and protracted disputes. The older leaders accused the younger people of styles of singing and dancing that were unseemly and immoral. They accused the leaders of this group of corrupting other young people morally. In retaliation, the younger people accused the older leaders of using their sermons to launch personal attacks and for preaching without basing their argument in the church's accepted texts. The older leaders had also been in receipt of salaries for a long time, and had invested in businesses in Konso. The younger group made serious accusations against the older leaders of financial and administrative corruption. They complained that businesses that the older leaders had set up, which in one case included a bar employing 'bar girls' and selling *tej* (honey wine), were inappropriate.

The argument was heated and was much more than a minor internal church or liturgical dispute. It potentially had serious repercussions on people's employment and livelihoods. In Karate town, two factions in the church developed: this led to conflict which became violent as fighting and stone throwing broke out on more than one occasion. Finally, they split into two distinct factions, worshipping in different locations; ultimately they became two separate churches. In addition, the Ethiopian Orthodox Church was becoming more powerful, and new Protestant Christian Churches had also come to the area and started to attract followers.

All of these developments weakened the moral authority and attractiveness of the *Mekana Yesus* church. At the beginning of the twenty-first century, it no longer enjoyed a privileged position as the major ritual and practical alternative to the customary networks. The number of members declined, especially in the villages. In Buso, when I asked a church leader why the numbers had declined so rapidly and extensively, he explained:

> It is difficult to know the reason. Some people have left and have changed back to the cultural life... One thing is that now there are many churches... The people believe that they give people money to come to their religion. Some give 50 *birr*, some give 100 *birr*. It is not like this here in this village, but I have heard that it is so elsewhere. When the people hear these things they say 'if people can change their beliefs for centimes, then this is no better than cultural life.' And so they have left Christianity and they have gone back to the traditional practices.[7]

In addition, he said, people are asking 'why should they follow people who tell you not to fight, but who are fighting all the time?' The *Mekana Yesus* church had been popular and strong when it provided spiritual and material forms of support and the promise of a better life. In the years that followed, its influence waned. As well as the fighting in the church, evangelizers were no longer at work because of the factional disputes. At the same time, there were fewer or no foreign missionaries[8] and *Mekana Yesus* was no longer the main provider of services and assistance. The government had improved its health care and provided more schools and health stations. NGOs have also been active supplying food relief and other forms of development assistance. The Protestant Christian church had also been through the 'honeymoon' period that it experienced at the beginning of the post-1991 period. Like other ideologies and movements, it had its own 'lifecycle' and was now entering a period in which its promises were no longer so fresh and exciting; there was a growing realization that the utopian promises were as remote as ever. The combination of these developments led many to become disillusioned, in the same way as they had been previously under the *Derg*. In addition, the post-1991 government's emphasis on democracy, on the right of everyone to live as they wished, and on the legitimacy of local 'culture', had changed the context in which *Mekana Yesus* operated. Local culture and local institutions were no longer necessarily considered inferior.

The revival of *poqallas* and other indigenous institutions

In the post-1991 period, the government classified indigenous institutions and cultural practices as acceptable. Possessing a distinct set of cultural institutions and practices was even advantageous. As the country was restructured along ethnic lines, origin myths, cultural practices, and linguistic criteria were used to place one group within the administrative boundary of another, or, as in the Konso case, give it a unique, independent status. Being a special *wereda* brought certain administrative challenges to those who had the new responsibilities, but it also brought recognition and new advantages. State funding was not funnelled through the intermediate stage of the zone, where it would be divided and could be diverted to others. Some of those in official positions in the special *wereda* also had the same authority, and wages, as those in another area at the zonal level. The distinctiveness of a particular region could be emphasized as a way to obtain such status.

The previous chapter has shown how many *poqalla*s managed to reassert their claims to land during the latter years of the *Derg*. In the post-1991 period, they started to revive their ritual and political role. This process was not without contestation, however. The Protestant Christians' equation of *poqalla* practices with devil worship had left a powerful legacy. Government officials might have wanted to promote Konso culture in theory, but in practice they were unsure about whether or not they wanted to extend this notion of 'culture' to include these particular indigenous institutions. For many, as I will show later, the *poqalla*s were still old landlords and symbols of exploitation; they were also a potential source of political competition. Any revival of the *poqalla*s that took place at this time was the outcome of a discursive struggle over legitimate and desirable rights and practices.

Soon after the change of government, the discursive struggle was expressed most vividly in an event that supposedly took place one market day. News of this event spread like wildfire. I heard it first from the father of the family I lived with when he came back from market in Gummayde. He said that Konso people had been sitting drinking at the market when a cockerel came among them and started to talk. It strutted among the people sitting there and warned them:

> This culture of ours of long ago, of the big families, you have destroyed it. Before, people respected the *poqallas*. Every year, they slaughtered a sheep and put a small piece of the skin on their wrists and they ate food mixed with butter. They called all their relatives together and they prayed to the great God for rain, and it rained. What has happened to these things nowadays? Now people come to your house and say these things are of Satan, and they go to the house and pull out all their cultural things.
>
> So today, watch me and see what I will do. Today people are dying, and in the evening they will die, and tomorrow and the day after tomorrow. This will continue until you respect the big families again.[9]

The story is an example of a narrative device that makes a case for a particular truth (Fortmann, 1995). It is not material as to whether or not it really happened. Wilder things are talked about in the market places and houses where *chagga* is sold. The point is to explore the work that the story performs. The story engages directly with the Protestant Christian valuation of matters related to the *poqallas* as devil worship, and it challenges that valuation. It is also literally crowing about the powers that *poqallas* have, and calling for them to be respected again. It highlights the way the *poqallas* are perceived as a means of mediating – ritually and practically – against disease and disaster, and warns of what will happen if *poqallas* continue to be ignored. The story reflected the general feeling that the customary practices and beliefs were plausible again.

The reinstating of the *poqallas* started slowly and had chequered outcomes. In Buso village, one of the first developments was the re-establishment of the position of *apa timba,* the main authority in the village. Disputes that could not be resolved at a more local level (for example the clan, lineage or neighbourhood) were referred to the *apa timba* for decision. The position of *apa timba* was symbolized by a drum, and the position itself circled in a fixed order between prominent families, most of whom were also *poqallas*. In Chapter Four, I showed how the *apa timba* is a form of power often held by the *poqalla* and which augments his position. During the *Derg* years, the drum in Buso was hidden, in order to avoid it being confiscated by the *Derg,* and while the drum was hidden the position stopped functioning.

After the change of government, the drum was taken from its hiding place, dusted down, mended where necessary, and circled again. Initially, it soon came to a halt as one man, a *poqalla,* who was due to accept the drum, refused it. Although he was not a Protestant Christian himself, many members of his immediate family were, and they put pressure on him not to accept it, on the grounds that the responsibilities included a sacred, but non-Christian dimension. In the meantime, the person who had been holding the office became frustrated with the delay. He cut his hair (a taboo for the *apa timba*) and said 'I have done my duty and I will not tolerate these head-lice any longer!' The circling of the drum stalled, but soon after, it passed to another and, with the declining influence of the Protestant Christians, it began to function again. By 2002, it was well reestablished.

Elsewhere, other attempts were taking place formally to revive certain *poqallas* by carrying out key rituals which had been left undone under the previous regimes. The family of Irmale *poqalla* from Fasha region, for example, demonstrates the way in which some *poqallas* were literally 'unearthed' during this period. The family exhumed the bodies of two generations of *poqallas*, father and son. Dowate Irmale, the father, had been buried 26 years previously, and Palsha Irmale, the son, had been buried for only three years. Dowate Irmale was the last Irmale *poqalla* to have been inaugurated properly and no new *poqalla* could be inaugurated until Dowate, and his son Palsha, had been buried and mourned correctly.

The bodies were exhumed and a man named Shola Tsomona was given the responsibility of reconstructing the bones into a form that could be recognized as Dowate and Palsha. He had photographs of both men in order to assist him in this task. First, he cleaned the bones, a more difficult process in the case of the second body that had only been buried for three years. Then, most of the bones were placed in a sack with the skulls on top. The skin sack was covered in a Konso blanket, so that it looked as if the person was sitting on a chilly day, with a blanket drawn around him. With the help of two other men, Shola heated a black substance,[10] which, when heated becomes sticky and pliable. Using this substance, Shola reconstructed the faces on the skulls and on to the new faces he placed eyes and teeth made from animal bone. Finally, striped cloths of a kind favoured by men during ceremonies were placed on the tops of their heads, and *xallashas* were placed on their foreheads.

When the reconstruction was complete, the two bodies sat displayed proudly with their photographs in frames beside them (see Photos 7.1–7.3). Until these bodies were reburied, as they were to be two days later, they received visitors. This is similar to the ritual that surrounds the most important regional *poqallas* after their deaths: the bodies are disembowelled, the eyes and the brain are removed and the body is washed with a mixture of honey, water and various plants, in order to preserve it. The body is then kept in a certain house for a period of time during which it is said that the *poqalla* 'is unwell' or 'has a cold' (Tadesse Wolde, 1992). This time can last for months or years, and people visit, bring gifts, and pay their respects to the *poqalla*, but mourning is forbidden. The exhumed bodies of the Irmale father and son had a similar status. No one was allowed to mourn their deaths until they had been buried with full ceremony. People entered the house containing the reconstructed bodies and paid their respects. The ritual was evidently serious, but this did not prevent people from enjoying it. When the first body was finished, I wrote in my notebook that 'it sat amongst us and people talked to him,

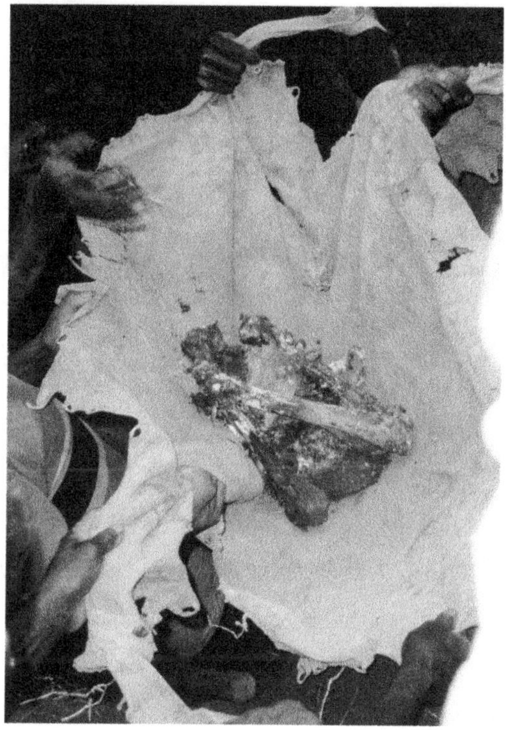

Photo 7.1 *Some of the bones of the* poqallas *as they were taken from the ground*

commiserating with him for having a cold, offering him *chagga* (beer) and mock wiping his nose.' When the second body was finished and it was time for me to pay my respects by greeting him and asking him how he was, someone slipped their hand behind the body so that it was as if he could shake my hand. When the body of the older *poqalla* seemed to be slumping, someone said to him 'I can see that you don't feel well, it is no wonder – you are getting so old!'

Finally, the two bodies were reburied, and everyone danced in a celebration called *man'na*. Eighteen animals were slaughtered, nine for each man, and the *poqallas* were deemed to have been buried correctly. Following these rituals, the new Irmale *poqalla* could be inaugurated.

The Irmale family ritual demonstrates the extent and the expense of the ritual processes which accompany the inauguration of the *poqalla*. Obtaining eighteen animals for slaughter, especially eighteen bulls, is an enormous undertaking. It also shows that the *poqallas*, as an institution, can, literally, lie dormant until the family has access to the resources to carry out the ceremony, or until the social and political climate changes to one in which the rituals are more acceptable. In Irmale's case, the two *poqallas* appeared to have lain dormant for a combination of these reasons. This dimension of the *poqalla*'s position brings a particular degree of

Photo 7.2 *Shola and his assistant giving the older Irmale a drink*

Photo 7.3 *With both* poqallas *after they have been reconstructed*

resilience. They are able to be 'reawakened' after doing nothing for a very long period of time.

Renegotiating power and position

The reestablishment of the *poqallas* has been progressive but, in the same way as its start was contested, the ongoing process has required continual negotiation. As discussed, under the *Derg*, some *poqallas* had already started secretly to reclaim land that they had lent to others, and to reclaim labour of those who had borrowed that land. In the post-1991 period, some *poqallas* have attempted to reinstate more publicly other aspects of their role, including the juridical, social and ritual. How this has been done has varied from village to village, region to region, and *poqalla* to *poqalla*. It is not possible for *poqallas* to exist and operate as they did in the pre-*Derg* or pre-Imperial situation. Those who wish to take a public role again have had to renegotiate their power and position.

Overall, the *poqallas*' claims to power have been weakened by the history discussed in the previous chapters. Economically, they have not managed to reclaim access to all the land or labour that they had before. When those who have the *poqalla*'s land go to labour for him, the labour that is given is not, as before, called *piyolada* and given systematically three times a year; it is given much less frequently and only in return for good strong *chagga*. It is possible also that, as population growth has taken place, landholdings have decreased, further reducing the *poqallas*' control over this important form of capital. Juridically, the position of *apa timba* may have been restarted in Buso, but it seems unlikely that the *poqallas* will be called to judge as often as they were before. Now, village residents often turn to the Chairperson of the Peasant Association when they have a grievance. Ritually, some *poqallas* may have been inaugurated again, but few seem to have the wealth, inclination, or pressure from their lineage, to perform rituals regularly, if at all. The *poqallas* with small lineages are 'weak like children', and unlikely to be reinstated (see Chapter Three). Although many *poqallas* have been revived, their position is only a part of what it was before.

As former claims to power have grown weaker, *poqallas* have sought new ones. They have tried, with greater or lesser levels of success, to gain access to new forms of prestige and power. This can be seen at a village and a regional level. At a village level, in Buso, for example, it is notable how many *poqallas* are sitting as elected members on the new village decision making body, referred to locally as the *parlamma*.[11] The *parlamma* is an adaptation of the Peasant Association Committee set up under the *Derg*. It fulfils very much the same functions. The group meets monthly to review the work of a smaller executive group who manage the village day to day.

There are more than just *poqallas* sitting on the *parlamma* in Buso, but the election of several young *poqallas* onto this decision-making body is significant. From one perspective, it could be seen as a positive development, as it represents a convergence between, on the one hand, the institutions of state, which have expressed the intention to govern and to generate development in a way which is both based in, and appropriate to, local culture; and, on the other hand, the indigenous institutions who have historically been central to that local culture. Whereas, previously, the state and the *poqalla*s were opposed, this arrangement represents an alliance. Such a convergence would seem to have the capacity to improve participation, and to encourage the genuine construction of a new Konso state, embedded in local culture (Watson, 2003). It would appear therefore to meet some of the goals set out by the new ethnically federalized state. From the other perspective, by sitting on the *parlamma*, the *poqalla*s become members of the lowest rung of the state administrative structure, and intermediaries between the government and the people. It is not difficult to see parallels between this situation, and the one that existed under northern rule, when some *poqalla*s acted as *balabbat*s and were intermediaries between the state and the people. Some of the *balabbat*s (not all) received benefits in return for their services, and received new powers and responsibilities that changed their position, their power, and their relations with other people. It has yet to be seen whether or not certain *poqalla*s use their new position in this administrative structure to their own advantage. It is too early to judge the way in which they are viewed by other people, and whether or not they are considered legitimate.

At a regional level, the complexities that accompany attempts to renegotiate power and status become more evident. Here a rather different set of power relations between the *poqalla*s and the current state regime emerges. One regional *poqalla* has been relatively successful at repositioning himself and claiming power, but he has also experienced serious problems as a result. The Konso administration and some of his neighbours have viewed his actions with suspicion, and the struggle has become overtly political.

The regional *poqalla* in question has been practising some of his customary rituals and has also been involved in peacemaking. These activities have brought him a strong reputation and the respect of people far from his customary territory. Problems have emerged, however, as he has attempted to reassert his claims to wide lands that are being used by others. Whereas in the villages, many *poqalla*s were able to reclaim land even during the *Derg* years, it has proved more difficult for the regional *poqalla*. As he lives apart from the villages, his relations with those using the land are more distant; the land he is claiming is also more extensive. Disquiet has arisen recently as, according to disgruntled neighbours, the regional *poqalla* has attempted to use the modified land law to justify his claims. The new land law claims that leasing, sharecropping and mortgaging of land are permitted. Hence, whereas at a village level, accepting a *poqalla*'s claims to land means helping the *poqalla* occasionally when he calls for assistance on his fields, accepting the regional *poqalla*'s claims to land, framed by the new land law, potentially has much more serious repercussions: a rent, mortgage fee, or share of the crop, could be requested. The *poqalla*'s recourse to state law to support his customary claims has been largely viewed as unfair. One man, for example, when asked what he thought about the *poqalla*'s actions, commented: 'You seen now some people are owning wide lands. And in the future it may turn back to how it was in Menelik's and Haile Selassie's time.'[12] The *poqalla*'s claims have thus not been seen as having the legitimacy of custom, or the 'eternal yesterday'. On the contrary, they have been constructed as imperialist.

The situation has been compounded as the *poqalla* in question has become frustrated with the lack of recognition of his claims, and he has declared sympathy for the political opposition to the ruling EPRDF. His attraction to the opposition party has been largely based on their support for private land tenure; if the opposition party had won, he felt his extensive landholdings would have been secured. Declaring sympathy for the opposition party was a bold move, however. In Konso, members of the opposition party have been viewed with suspicion and hostility by the current administration.

The problems experienced by this *poqalla* were rather atypical. The example has been included because it demonstrates that the revival of the power and roles of the *poqalla* is not always straightforward. The *poqallas* are claiming power and position in a new context where their actions have new significance. For their position and power to be reinstated successfully, it must be considered legitimate by those among whom they live. But the example is also of wider significance for development, and more specifically for attempts by development organizations to form partnerships with indigenous institutions.

In Konso, there is a new NGO that aims to work with elders and ritual leaders in the hope of fostering community action for development and encouraging environmental protection. Known as Parka, from the name of the Konso work group, it is an example of a development organization that has attempted to work with indigenous institutions to facilitate development and improved environments. Parka has done some excellent work, but it has also experienced problems because of its attempts to work with indigenous institutions. In this case, Parka's association with the regional *poqalla* discussed above, led at times to accusations that the organization is imperialist and/or supporting the opposition. These accusations caused the Konso administration to close the organization for a period, and only to reopen it after appeals were made by the organization's representatives to the Central Government.

The politics of forming partnerships with indigenous institutions are apparent and often problematic. It may still be desirable to work with indigenous institutions, as they *may* be legitimate cultural representatives. But the point is, there is a politics inherent in these institutions which cannot be ignored. The politics affect relations within the 'community' (discussed earlier), but also relations between a 'community' and any development organization wishing to cooperate with them, and others further afield. Development organizations may be currently hoping to engage more proactively with 'culture', but this example shows that engagement with culture and with key leaders is not neutral.

New context, new challenges

In the twenty-first century, the political and economic context in which the *poqalla*s find themselves has radically changed in other ways. More specifically, the expansion of the market has brought certain other limitations to the extent to which *poqalla*s can reinstate their power and position. The market has undermined the political ecology of the orthodoxy that was based in control over land. There are also further new discourses and networks of trust. In this final section, I explore these new changes very briefly through the example of the relationship between farmers and traders in Konso.

Artisanal traders (*xawuda*) have historically been marginalized from the structures of power in Konso. They had no land, and did not participate in the much valued agriculture, and were considered to be at the bottom of a stratified society in which the large landholders, the *poqalla*s, were at the top. In a society that valued the 'first comer', and men above women, the artisanal traders were referred to as 'little brothers' or 'sisters', metaphors that reinforced their subordinate position (see Chapter Four).

Traders were initially artisans and many still depend for their living on blacksmithing, weaving, butchery, tanning and pottery. They also had their own formal network of traders called *fuld'o*. This institution has its headquarters in Dokato, in Konso, but the network stretches from Moyale to Addis Ababa. The network has its own rituals, its own principles and its own codes of practice. It supports the activities of its traders, and imposes fines and sanctions on those who cheat others in the network (Amborn, forthcoming 2009).

As Konso has been incorporated further into national and international markets, the arrival of manufactured and often foreign goods has led to a decline in the demand for the *xawuda*'s home produced crafts. The markets have filled with secondhand clothes exported from Europe and America, and cheap new clothes from China and other parts of South East Asia, and the demand for cloth woven in Konso has declined. Tools, utensils and plastic bowls made in factories have replaced the tools and pots made by the artisans. Thus, artisanal traders have had to rely more on trading than on artisanal production. This created difficulties initially, but, ultimately, many of the artisanal traders have managed to make good use of the new market opportunities available, and some have thrived.

The artisans' success has also come because they formed a powerful alliance with other traders in the country. Northern traders whose families came to Konso in the time of the Empire have been particularly central here. Many of these northern traders have been invited to join the *fuld'o* network. At the same time, many Konso artisans have joined the Ethiopian Orthodox Church, the main religion of the northern traders. Although new converts were probably inspired to convert for spiritual and religious reasons, joining the church undoubtedly had other advantages. As one new convert explained:

> When traders go on the road and trading, the people ask, 'What is your religion?' When you say 'I am Orthodox' then you are treated very well. In Moyale, Arba Minch, and other places there are Orthodox churches. If you have any problems you can say, 'I am Orthodox' and they will accept you in everything.[13]

Previous chapters have shown how ritual meaning and religious group membership (customary or Protestant) were important in fostering co-operation and structuring economic activities. Here, membership of the Ethiopian Orthodox Church and the *fuld'o* network provide another powerful and meaningful form of social capital structuring economic activity. In this case, the activity is market-based trading. In 1995-6, the number of Konso members of the Ethiopian Orthodox Church was few. By 2002, there were several new churches and membership was strong and growing. The growth appears to follow the expansion of trade activities in Konso. The Ethiopian Orthodox Church can therefore be seen as a new force and a new discourse in Konso, structuring economic activities and encouraging forms of cooperation.

These developments are indicative of wider processes of change taking place in Konso. Older identities and antagonisms are being restructured. In the past, Konso livelihoods were based in agriculture, and power was previously based in control over land. Artisanal activities and trading were unrewarding minority activities. As artisanal activities have started to pay more, now many people have diversified from agriculture. Some Konso farmers say 'We are all *xawuda* (artisans) now.' As a consequence, agricultural activities, and the structures of power that accompanied them, specifically those of the *poqalla*s, are no longer so strong or so central to Konso lives and livelihoods. In Chapter Three, I explored how the *poqalla* was at the apex of a set of unequal power relations. The *poqalla*'s relations with his lineage mirrored and legitimized unequal relations at other levels and scales. It is possible that the transformation that has taken place in the position of the *xawuda* may reflect changes to other power relations at other scales. In this new context, the position and power of the *poqalla*, which was based in control over land and its use to obtain access to others' labour, is unlikely to be revived fully to the extent that it was before.

Notes

1 In practice the EPRDF has discouraged any movement that they have seen as a move towards secession on the part of any of the constituent ethnic groups (Young, 1996; Fullerton Joireman, 1997).

2 The Millennium Development Goals were adopted by the member states of the United Nations in 2000. They set out a series of development objectives to be achieved by all countries by 2015 (for more details see http://www.developmentgoals.org/; accessed 14 April 2004).

3 Killano, 1996.

4 See Watson, 2002 for more details, and Kurimoto, 2002, for a comparison with the Anywaa experience.

5 It was hoped that some of the education could be carried out in the vernacular language also.

6 These rights were enshrined in the Constitution as 'the right to freedom of thought, conscience and religion. This right shall include the freedom to hold or to adopt a religion or belief of his choice, and the freedom, either individually or in community with others, and in public or private, to manifest his religion or belief in worship, observance, practice and teaching' (FDRE, 1995: Article, 27(1), 8-9).

7 Interview, K. K., Buso, 2002.

8 I do not want to exaggerate the importance of foreign missionaries but I think they have been symbolically important, may have brought access to certain resources and contacts with a world further afield, and may also have had a calming influence on the local disputes.

9 Xalale Xawde, 1996.

10 Called, in *Afa Xonso*, *tankarita*. It is made from the sap of a tree and it is also used as a kind of incense.

11 Within the *parlamma* is a smaller executive group referred to as the *cabinet*. These new terms reflect the use of the ideas of 'parliament' and 'cabinet' as models of how the new democratic politics in Ethiopia should work, from the centre down to the village level. The presence of these terms at the local level also reflects the degree to which these ideals have been embraced and internalised.

12 Neighbour, 2002.

13 Qaarta A., 2002.

Conclusion

Landscape, Meaning & Development

The last three chapters focused on the changes that have unfolded in Konso, and the way they have impacted on the institutions that are central to the production of Konso indigenous agriculture and landscape. Together with the earlier chapters, they emphasize the way in which material struggles over land and labour are played out in discursive struggles over the legitimacy of different beliefs and forms of identification (Konso 'custom', Protestant Christian, Orthodox Christian). In this conclusion, I return more explicitly to the landscape and examine what can be learnt from the material presented here about how landscapes like Konso are produced and maintained over time. I also explore the significance of this account for development practice and policy.

The principal message is that the terraced landscape of Konso is a cultural landscape; it is produced out of networks of collaborative action which are also networks of meaning. Historically, the indigenous institution of the *poqalla* has been central to the network of meaning, and has helped to produce the landscape and the indigenous intensive landscape that it supports. The *poqallas* have been powerful institutions because they were intimately connected to others in Konso, who identified with them personally: the *poqallas* were their relatives; they represented, structurally, every Konso person's father and their origins. The *poqallas* also performed multiple functions, providing material, juridical, social and cultural support. In rituals, the *poqallas* were thought to be able to influence forces beyond the control of ordinary mortals. By participating in the network of ritual and practical activity, individuals could hope to secure a more prosperous life, one in which the fertility of land, themselves and their animals was assured. The institution of the *poqalla*, which mobilized labour and applied it in the construction of landesque capital, was connected to all aspects of people's lives and their wellbeing. The relationship between a *poqalla* and others was practical, personal, emotional and ritual. The work through which the landscape was constructed was similarly so.

Over time, other discourses and networks of collaboration have emerged in Konso and rivalled the discourse that surrounds and supports the *poqalla* (described in this book as the Konso orthodoxy). In the new modern and postmodern discourses, what is striking is that there continues to be a component of 'belief' that structures new networks, practices and ways of managing land, labour and other economic activity. In the *Derg* years, and the early years of the EPRDF government, Protestant Christian (*Mekana Yesus*) ideas and networks of cooperation proved important. During the EPRDF years, and as market opportunities and connections to other parts of Ethiopia have become stronger, membership of the Ethiopian Orthodox Church has become more significant. The ethnographic material presented in the book suggests that in establishing certain economic practices, especially in this case relating to the management of land and labour, people's frameworks of belief, their visions of themselves and their place in the world (this world and the one of *Waqa*/God and the spirits) are critically important.

Although new discourses of meaning and practice have emerged in recent years, the discourse and institution of the *poqalla*s has shown a remarkable level of resilience. Cultural change is often viewed as cultural breakdown, but this is not the case here. The Konso material demonstrates that there is an ongoing struggle to reassert the *poqallas*' position and power. The *poqalla*s still have relevance to people's lives and, in some ways, they have reclaimed their power and influence. However, the changes that have taken place in the last half of the twentieth and first decade of the twenty-first centuries have left their mark. The new context, with new roads and new economic opportunities, has meant that the *poqallas*' control over land and symbolic capital is now rivalled by others' control over land and other forms of capital; *poqalla*s are not as powerful as they once were.

The task at hand then becomes to address the question set out in the introduction of this book, and to explore whether the cultural changes described are evident in the landscape. If the landscape is produced from these different cultural institutions, then are changes to the cultural institutions visible, traced out on the landscape? As this book has shown, the answer is not an easy one: identifying a causal link between cultural change and landscape form is not straightforward. There are more variables at play than simply the *poqalla* and the degrees of intensification. Population growth has taken place, partly as a result of the eradication of the cultural institution of *farayta*, that limited birth rates (see Chapters Five and Six). Environmental change may also be a factor. The scale of a landscape study also complicates the investigation, as it is very difficult to examine changes to the extent and nature of intensification Konso-wide. There is also little historical data with which to compare contemporary data in order to say something about change over time. Conclusive study requires data which are yet to be collected.

But it is possible to posit some conclusions. Where the everyday management of the agricultural stone walled bench terraces, and the preparation of land for cultivation, the weeding, the sowing and harvesting are concerned, the work appears to be being carried out in similar ways to those which existed before, and equally effectively. Effort is still being applied to make the most of the land and water available. If old forms of cooperation have declined, new forms of cooperation appear to have arisen to fill their place. Membership of new labour networks may not depend on shared orientation to a particular *poqalla*, but on shared new religious beliefs and church membership. Therefore again, social and cultural change is not seen to lead automatically to social or environmental disaster; life still goes on, as people adapt and organize themselves in new and meaningful ways to meet the continual challenges of maintaining a living.

The situation is different where larger forms of landesque capital are concerned. Here, the social and cultural changes described in the book may be having a more visible impact on the landscape. In Buso village, for example, recent flooding has destroyed many of the large flood protection walls (*tomota*) and irrigation channels which enabled some of the most productive land to be cultivated. In the past, as has been seen, these large flood protection walls were often built by *poqalla*s, who had access to the fertile irrigated land, and to the labour that was required to maintain them. In 1995–6, one *poqalla*'s wife called several neighbourhoods to labour for her, paying them and brewing the necessary *chagga* (see Chapter Two). Another *poqalla* called on his lineage, and those to whom he had lent land, to labour for him to reconstruct his wall (Chapter Two). The problems with erosion caused by floods may have worsened in recent years, but it seems probable that the decline in the power, wealth and influence of the *poqalla*s means that such large earthworks have become less well maintained. Their construction is less likely. There are areas, therefore, where the landscape is not being used as intensively as it was before, as few have the ability to raise enough labour to maintain the larger earthworks, or forms of landesque capital.

Meaning for development

The material presented in this book has direct relevance for development policy in Konso and beyond. It demonstrates the importance of indigenous institutions for environmental management. The social capital that indigenous institutions embody is vital to the production and reproduction of the Konso intensive agriculture and other landscapes like it. At the same time, these indigenous institutions cannot easily be replicated in other places, since they are part and parcel of people's identities and life worlds. The politics inherent in the institutions (and their relations with others) also makes forming partnerships with indigenous institutions for development a complicated endeavour. When partnerships are formed for development, existing social inequalities may be entrenched. In partnerships with indigenous institutions, the actions of development organizations may also take on wider social, political or religious significance. This significance may lie beyond the development organization's control and may influence the way their actions are accepted, modified or rejected by those with whom they wish to work.

The material presented here also challenges the mainstream readings of Konso by development organizations, and throws into doubt some of the policy recommendations that accompany those readings. These mainstream views are largely pessimistic and 'Malthusian', presenting a picture in which the population of Konso has expanded to a point where it is causing environmental degradation. The indigenous intensive agriculture is seen as having reached its limits; the population has exceeded the carrying capacity of the land, and the agriculture is no longer viable.

The development organizations promoting these views have come to Konso mainly since the 1991 change of government. Some such as Farm Africa, a UK-based NGO, have been involved in long-term development projects. Other organizations have come to Konso periodically to carry out relief operations, as food shortage situations have developed in the area, just as they have in many other parts of southern Ethiopia and northern Kenya. A brief look at some of their literature and practices illustrates the way Malthusian ideas are integral to many of their views and policy formulations. For example, the UNDP Emergencies Unit for Ethiopia (UNEUE) in 2003 claimed that:

> malnutrition is chronic in Konso, a *wereda* that has received food aid for 29 years on an almost permanent basis... Very urgent is also the dissemination of family planning methods. In only 20 years, the population of Konso has increased from 60,000 to 250,000!.. in recent years their ability to cope with natural calamities has reduced dramatically. Reason for this is exponential population growth and *the unsustainable exploitation of natural resources...* (UNEUE, 2003: 9–11, my emphasis).

Farm Africa, which has done some valuable work in Konso, has also espoused similar ideas. In one of their brochures designed to promote support for their 'capacity building programme' in Konso, they included photographs of extreme gully erosion from Gazergiyo region of Konso, and commented that 'dramatic land degradation is the prominent feature of the *wereda*'. They continued:

> The incompatibility of the community effort with the *detonating* [sic] *environmental situation* and fast growing population has left the community in lurch [sic]. There is not enough land in Konso to enable all its families to build up and maintain satisfactory levels of reserves through farming alone (Farm Africa, Ethiopia Capacity Building Project, publicity brochure, no date, my emphasis).

These Malthusian ideas are alarmist and powerful. They have even been adopted by some Konso people, to the point where they have become 'a local tongue'.[1] They form a discourse or narrative that frames the 'problem', together with a set of causations, victims, perpetrators and solutions (Ferguson, 1990). Scholars have shown how the Malthusian narrative has often been applied erroneously to African environments (Leach and Mearns, 1996; Fairhead and Leach, 1996). It is often held to be common sense and self-evidently true; the data on which it is based are often shaky or missing altogether. In Konso, there are also problems with the evidence underlying the model. Population size may have increased but the extent of population growth is difficult to assess accurately. Figures suggest that population growth has indeed been exponential, with estimates of 55-60,000[2] in 1972, 125,000 in 1986,[3] and 215,000 in 2002.[4] But each of these figures uses a different definition of Konso, and some may be more 'guesstimates' than 'estimates'. The UNEUE claims that food distributions have been carried out for 29 years continuously may sound authoritative, but the lack of records on such matters means that the history and scale of these distributions is obscure. Most significant for this research, the majority of the reports imply that land is becoming degraded and that the soil is becoming exhausted through overuse, and yet there has been little research into soil quality and soil degradation. The Farm Africa brochure shows pictures of 'dramatic' erosion, supported by a claim that there is extreme population growth and environmental degradation, implying that the catastrophic erosion in the picture is a result of these processes. But the area of erosion pictured is that of Gazergiyo, an area of Konso that has been eroded for a long time. It is the area of erosion mentioned in the Guufa *poqalla*'s origin myth, when he was invited to Konso to come and 'heal' the land. He placed his feet in the feet of elephants and walked over the land, and much of it healed, but he kept one area of the country – what is Gazergiyo today – as a reminder to the people of what he had done for them. Nowadays tourists and wedding parties go to Gazergiyo to admire the patterns of erosion, which they name 'New York' because the eroded landscape is thought to resemble skyscrapers: it is not a case of land eroded through agricultural overuse (see Photo 8.1).

It is not asserted here that there are no problems in Konso, or that, in years of poor rainfall, food shortages are not serious. But it is argued that further detailed research into the demographic and environmental processes (and any links between them) is necessary before conclusive statements can be made, and before policies are designed on the basis of them. In addition, the research presented in this book has underlined the way that the Malthusian model is inadequate, not only because the data on which it is based are often problematic (as in this case), but also because it draws conclusions based on comparing absolute numbers of people and absolute amounts of environmental resources. It assumes that 'more people' equals 'more erosion'. The Konso material (like that of Tiffen *et al.*, 1994; Ostrom 1990) shows that what happens to the environment depends on the

Photo 8.1 *Area of eroded land near Gazergiyo, known locally as 'New York'*

institutions which govern how those resources are used, and by whom.

In Konso, if environmental degradation is taking place, it is not a homogeneous process, and it is necessary to disaggregate the way it may impact on different areas and different groups of people. Many people are still investing in the land and manuring, terracing, and constructing rainwater harvesting structures and so on. Some of these conservation practices are being carried out in new ways. If environmental degradation is taking place it is not so much directly linked to population growth, but to the institutional history, to the decline of the *poqallas*, and to the inability of people to raise enough labour to maintain the larger forms of landesque capital. The impact of population growth is more likely to be seen in smaller plots for some, increased landlessness and out-migration. All these may create new vulnerabilities which require development support, but they are not the same as environmental degradation.

These points are significant because the dominance of the Malthusian discourse in development accounts and public discourses has powerful consequences. It leads to the belief that Konso agriculture and landscape is in crisis, that it is on the point of irreversible environmental breakdown, or that it is a cultural landscape only of the past. It suggests that, like the indigenous agricultural terraces of Nyanga in Zimbabwe and Engaruka in Tanzania, the Konso terraces might soon be abandoned. In this discourse, Konso agriculture is a lost cause, with little potential for development.

The conclusions from this study are quite the opposite. Although it may be necessary to support new alternative livelihood opportunities, Konso agriculture is still the mainstay of many livelihoods and it also requires support. As pressure on natural resources is high and growing, the indigenous agriculture may be more valuable than ever, as it still provides a relatively productive and intensive use of available resources. The decline in the *poqallas*' roles has meant that, despite high population pressure, land may not be being used to its maximum at present. There are opportunities, therefore, for development organizations to support the institutional dimensions of environmental management, in order that resources in Konso continue to be used as intensively and creatively as before. If organizations (as institutions) could work with Konso people and mobilize them for collective action, they could improve the current construction of flood protection walls, irrigation channels and other larger soil and water conservation structures.

As discussed, there are many pitfalls for development organizations that attempt to form partnerships with existing indigenous institutions, and it is difficult to construct social capital or institutions from scratch. But neither is impossible. To be successful in their partnership with Konso institutions and Konso people, development organizations could learn from the historical success of the *poqallas*. Successful partnership depends on building close relations with, and understanding of, the Konso people. Vital to all the networks of labour organization discussed in this book has been a shared set of values and beliefs. For development organizations to be able to construct social capital and institutions that function, they require a meaningful vision, religious or otherwise,[5] that they can communicate and that people can believe in. The only vision being communicated at the moment is the Malthusian one and, if anything, this pessimistic discourse is in danger of discouraging people from continuing the hard work involved in investing in and improving their environment. Moreover, it has not yet been shown to be happening in Konso. An alternative vision is needed: one which is positive, and which sees that intensive agricultural practice is still possible. New forms of collective action could be developed around the vision of continuing the intensification, and improving Konso environments and livelihoods by, for example, delivering clean water and badly needed improved sanitation. Konso people would no doubt welcome such a vision. While development organizations could draw on some of the positive qualities exhibited by the *poqallas*, it is important not to idealize them. The *poqallas* were/are not perfect. Within their discourse were/are multiple structures of inequality. By mobilizing people to work together, development organizations could perform some of the same functions as *poqallas*; their challenge might be to do this without inculcating the same levels of inequality.

As a final note, I'd like to reflect on T.S. Eliot's *The Waste Land*, because the poem has come to my mind many times while writing this book. Although Eliot's poem defies simplistic interpretations and single readings, I have attempted an interpretation and reading of Konso in a way which highlights some resonances with the poem. Behind Eliot's poem is the myth of the Fisher King, who ensures the wellbeing of his people. This gives way through time to an 'Unreal City' of people without aim or direction. They are in a desert, a 'Waste Land', physically, personally and spiritually. This metaphor is similar to the ways in which the Konso orthodoxy has structured its social, cultural and material world. Rituals carried out by the *poqalla*, who also kept moral codes, protected the fertility of Konso people and their world, personally and physically. If these rituals were not respected, then personal and environmental disaster would ensue. Now, the rituals are not so carefully respected and, in some ways, the environment is not being maintained as it was before. The certainty provided by the *poqallas*, personally, economically, environmentally and ritually, has given way to a time of uncertainty. The 'Waste Land', predicted by the Konso orthodoxy, appears to be in danger of coming to pass.

However, in Konso, it is more likely that there never was a fully functioning closed loop between Konso culture and the environment, or a perfect fit between the *poqallas*' social and ritual acts and environmental management that was accepted by and participated in by everyone. The personal security of the orthodoxy was probably more ideal than real. The *poqallas* probably always were challenged, by their clientele who took too much, or by cults similar to that of *ayana* spirit possession, which provided alternative frameworks of meaning. The loss of the *poqallas*' power has not yet given way to complete disaster. On the contrary, only some areas are going uncultivated; Konso people have been shown also to make and remake meanings, from the old and the new, and to use these to structure collective action. The 'Waste Land', like the Malthusian narrative, is far from inevitable; even in Eliot's poem there are signs of renewal. The challenge is for people – in Konso and beyond, in partnership – to continue to generate new ideas and meanings. If those new ideas and meanings are believable, and resonate with people's experiences and interests, then they can galvanize people to work together to use the existing resources as effectively, efficiently and creatively as possible.

Notes

1 This phrase is Ferguson's (1999: 84). He uses it to describe modernization theory as promoted by local people in Zambia.

2 Hallpike, 1972.

3 Government Census.

4 Ministry of Agriculture, Konso.

5 The religious discourses have proved particularly powerful in the past, but shared secular ideas and values may have certain other advantages over the religious ones.

References

Adams, W. M. 1992. *Wasting the Rain: Rivers, People and Planning in Africa*, London: Earthscan.

Agarwal, A. 2001. 'Common Property Institutions and Sustainable Governance of Resources', *World Development*, 29 (1): 1649–72.

Amanor, K.S. 2002. 'Indigenous Knowledge in Space and Time', in H. Brookfield, C. Padoch, H. Parsons and M. Stocking (eds), *Cultivating Biodiversity: Understanding, Analysing and Using Agricultural Diversity*, London: ITDG Publishing, pp. 126–32.

Amborn, H. 1988. 'History of Events and Internal Development: the Example of the Burji-Konso Cluster, in Tadesse Beyene (ed.), *Proceedings of the Eighth International Conference of Ethiopian Studies in Addis Ababa*. Addis Ababa: Institute of Ethiopian Studies, pp. 751–67.

——. 1989. 'Agricultural Intensification in the Burji-Konso Cluster of South-West Ethiopia', *Azania*, XXIV: 71–83.

——. 2009 forthcoming. 'Mobility, Knowledge and Power: Craftsmen in the Borderland', in G. Schlee and E.E. Watson (eds), *Changing Identifications and Alliances in North-East Africa, Volume 1*. Oxford: Berghahn.

Anderson, B. 1991. *Imagined Communities: Reflections on the Origin and Spread of Nationalism*, London: Verso Books.

Anderson, D. 1984. 'Depression, Dustbowl, Demography and Drought: The Colonial State and Soil Conservation in East Africa during the 1930s', *African Affairs*, 83 (332), 321–43.

Appadurai, A. 1986. 'Introduction: Commodities and the Politics of Value', in A. Appadurai (ed.), *The Social Life of Things: Commodities in Cultural Perspective*, Cambridge: Cambridge University Press, pp. 3–63.

Arén, G. 1978. *Evangelical Pioneers in Ethiopia: Origins of the Evangelical Church Mekane Yesus*. Stockholm: EFS Förlaget.

——. 1999. *Envoys of the Gospel in Ethiopia: In the Steps of the Evangelical Pioneers, 1898–1936*, Stockholm: EFS Förlaget.

Bahru Zewde, 1991. *A History of Modern Ethiopia, 1855–1974*, London: James Currey.

Barnes, T.J. and Duncan, J. (eds). 1992. *Writing Worlds: Discourse, Text and Metaphor in the Representation of Landscape*, London: Routledge.

Bassett, T. J. 1993. 'Introduction: The Land Question and Agricultural Transformation in Sub-Saharan Africa', in T.J. Bassett and D.E. Crummey (eds) *Land in African Agrarian Systems*, Madison, Wisconsin: University of Wisconsin Press, pp. 3–31.

Baxter, P.T.W. 1978. 'Boran Age-sets and Generation-sets: *Gada* a Puzzle or a Maze?', in P.T.W. Baxter and U. Almagor (eds), *Age, Generation and Time: Some Features of East African Age Organization*, London: C. Hurst, pp. 151–82.

Bayart, J-F. 1993. *The State in Africa: The Politics of the Belly*, London: Longman.

Bebbington, A. 1997. 'Social Capital and Rural Intensification: Local Organizations and Islands of Sustainability in the Rural Andes'. *Geographical Journal*, 163 (2): 189–97.

Behrend, H. and Luig, U. (eds), 1999. *Spirit Possession, Modernity and Power in Africa*, Oxford: James Currey.

Benko, G. 1997. 'Introduction: Modernity, Postmodernity and the Social Sciences', in G. Benko and U. Strohmayer (eds), *Space and Social Theory*, Oxford: Blackwell Publishers, pp. 1–48.

Berman, M. 1982. *All That is Solid Melts into Air: The Experience of Modernity*, London: Verso.

Berry, S. 1989a. 'Access, Control and Use of Resources in African Agriculture: An Introduction'. *Africa*, 59 (1): 1–5.

——. 1989b. 'Social Institutions and Access to Resources'. *Africa*, 59 (1): 41–55.

——. 1992. 'Hegemony on a Shoestring: Indirect Rule and Access to Agricultural Land'. *Africa*, 62 (3): 327–55.

Black, P. and Shako Otto, 1973. 'Konso Dictionary'. Unpublished manuscript.

Black, R. and Watson, E.E. 2006. 'Local Community, Legitimacy, and Cultural Authenticity in Postconflict Natural Resource Management: Ethiopia and Mozambique', *Environment and Planning D: Society and Space*, 24 (2): 263–82.

Blaikie, P. and Brookfield, H. 1987. *Land Degradation and Society*, London: Methuen.

Blakely, T.D., van Beek, W.E.A and Thomson, D.L. (eds), 1994. *Religion in Africa: Experience and Expression*. Oxford: James Currey.

Boddy, J. 1995. 'Managing Tradition: "Superstition" and the Making of National Identity among Sudanese Women Refugees', in W. James (ed.), *The Pursuit of Certainty: Religious and Cultural Formulations*. London: Routledge.

Boserup, E. 1965. *The Conditions of Economic Growth: The Economics of Agrarian Change Under Population Pressure*, Chicago: Aldine.

Bourdieu, P. 1977. *Outline of a Theory of Practice*, Cambridge: Cambridge University Press.

——. 1984. *Distinction: A Social Critique of the Judgement of Taste*, London: Routledge & Kegan Paul.

Brøgger, J. 1986. *Belief and Experience among the Sidamo: A Case Study Towards an Anthropology of Knowledge*, Oslo: Norwegian University Press.

Bruce, J.W. 1990. *Community Forestry: Rapid Appraisal of Tree and Land Tenure*, Rome: FAO.

Brumfiel, E.M. and Earle, T.K. 1987. *Specialization, Exchange and Complex Societies*, Cambridge: Cambridge University Press.

Busk, D. L. 1957. *The Fountain of the Sun: Unfinished Journeys in Ethiopia and the Ruwenzori*, London: M. Parrish.

Butler, J. 1990. *Gender Trouble*, London: Routledge.

Carswell, G. 2007. *Cultivating Success in Uganda: Kigezi Farmers and Colonial Policies*, Oxford: James Currey.

Chabal, P. and Daloz, J-P. 1999. *Africa Works: Disorder as Political Instrument*, Oxford:

James Currey.

Chasseguet-Smirgel, J. 1983. 'Perversion and the Universal Law', *International Review of Psycho-Analysis*, 10: 293-301.

Chayanov, A. V. 1966. *The Theory of Peasant Economy*, edited and translated by Daniel Thorner, Balile Kerblay, and R. E. F. Smith. Homewood, Illinois: Richard and D. Irwin.

Clapham, C. 1982. 'Clientelism and the State', in C. Clapham (ed.), *Private Patronage and Public Power: Political Clientelism in the Modern State*, London: Frances Pinter, pp. 1–36.

——. 1988. *Transformation and Continuity in Revolutionary Ethiopia*, Cambridge: Cambridge University Press.

——. 1992. 'The Socialist Experience in Ethiopia and Its Demise', in A. Hughes, (ed.) *Marxism's Retreat from Africa*. London: Frank Cass, pp. 105–25.

——. 2002. 'Controlling Space in Ethiopia', in W. James, D. Donham, E. Kurimoto and A. Triulzi (eds), *Remapping Ethiopia: Socialism and After*, Oxford: James Currey, pp. 9–32.

Cleaver, F. 2000. 'Moral Ecological Rationality, Institutions and the Management of Common Property Resources', *Development and Change*, 31(2): 361–84.

Clifford, J. 1986. 'Introduction: Partial Truths', in J. Clifford and G. Marcus (eds) *Writing Culture: The Poetics and Politics of Ethnography*, Berkeley: University of California Press, pp. 1–26.

Colson, E. 1971. 'The Impact of Colonialism on the Definition of Land Rights', in V.Turner (ed.), *Colonialism in Africa 1870–1960, Vol 3, Profiles of Change: African Society and Colonial Rule*. Cambridge, Cambridge University Press, pp. 193–215.

Comaroff, J. and Comaroff, J. (eds) 1993. *Modernity and its Malcontents: Ritual and Power in Postcolonial Africa*, Chicago: University of Chicago Press.

—— 1997. *Of Revelation and Revolution: The Dialectics of Modernity on a South African Frontier, Volume 2*, Chicago: University of Chicago Press.

Crewe, E. and Harrison, E. 1998. *Whose Development? An Ethnography of Aid*, London: Zed Books.

Darragon, L. 1898. 'Éthiopie: Le Sidama, L'Amara, le Konso, etc', *Comptes Rendus des Séances de la Société de Géographie et de la Commission Centrale*, Mars 4, 18 et 21: 137–40.

De Boef, W., Amanor, K.S., Wellard, K. and Bebbington, A. (eds) 1993. *Cultivating Knowledge: Genetic Diversity, Farmer Experimentation and Crop Research*, London: Intermediate Technology.

De Soto, H. 2000. *The Mystery of Capital: Why Capitalism Triumphs in the West and Fails Everywhere Else*, London: Bantam Press.

Delaney, C. 1991. *The Seed and the Soil: Gender and Cosmology in Turkish Village Society*, Berkeley: University of California Press.

Denevan, W. M. 2001. *Cultivated Landscapes of Native Amazonia and the Andes*, Oxford: Oxford University Press.

Dessalegn Rahmato, 1984. *Agrarian Reform in Ethiopia*, Uppsala: Scandinavian Institute of African Studies.

Diamond, J.M. 2005. *Collapse: How Societies Choose to Fail or Survive*, London: Allen Lane.

Donaldson Smith, A. 1897. *Through Unknown African Countries: The First Expedition from Somaliland to Lake Lamu*, London: E. Arnold.

Donham, D. L. 1986a. 'Old Abyssinia and the New Ethiopian Empire: Themes in Social History', in D.L. Donham and W. James (eds), *The Southern Marches of Imperial Ethiopia: Essays in History and Social Anthropology*, Cambridge: Cambridge

University Press, pp. 3–48.
——. 1986b. 'From Ritual Kings to Ethiopian Landlords in Maale', in D.L. Donham and W. James (eds), *The Southern Marches of Imperial Ethiopia: Essays in History and Social Anthropology*, Cambridge: Cambridge University Press, pp. 69–95.
——. 1990. *Work, Power, Ideology: Central Issues in Marxism and Anthropology*, Cambridge: Cambridge University Press.
——. 1999. *Marxist Modern: an Ethnographic History of the Ethiopian Revolution*, Berkeley: University of California Press; Oxford: James Currey.
——. and James, W. (eds) 1986. *The Southern Marches of Imperial Ethiopia: Essays in History and Social Anthropology*, Cambridge: Cambridge University Press.
Donkin, R.A. 1979. *Agricultural Terraces in the Aboriginal New World*, Tucson: University of Arizona Press.
Douglas, M. 1987. *How Institutions Think*, London: Routledge.
Dreyfus, H.L. and Rabinow, P. (eds), 1986. *Michel Foucault: Beyond Structuralism and Hermeneutics*, Brighton: The Harvester Press.
Dumont, L. 1980. *Homo Hierarchicus: The Caste System and Its Implications*, Chicago and London: University of Chicago Press.
Duncan, J.S. and Duncan, N.G. 2004. 'Culture Unbound', *Environment and Planning A*, 36: 391–403.
Durkheim, E. 1965 [1915]. *The Elementary Forms of Religious Life*, New York: Free Press.
Eagleton, T. 1996. *The Illusions of Postmodernism*, Oxford: Blackwell Publishers.
Ehret, C. 2002. *The Civilizations of Africa: a History to 1800*, Oxford: James Currey.
Eide, O.M. 2000. *Revolution and Religion in Ethiopia: 1974–85*, Oxford: James Currey.
Eliade, M. 1957. *The Sacred and the Profane: The Nature of Religion*, San Diego: Harcourt Brace and Company.
Escobar, A. 1995. *Encountering Development: The Making and Unmaking of the Third World*, Princeton: Princeton University Press.
Estiphanos Berishia, n.d. 'The Family of Bata' the High Priest', unpublished essay.
Fairhead, J. 1992. 'Indigenous Technical Knowledge and Natural Resources Management in Sub-Saharan Africa: A Critical Overview', unpublished research paper.
——. 2000. 'The Conflict over Natural and Environmental Resources' in E. Nafziger, S. F. Wayne, and R. Vayrynen, (eds), *War, Hunger and Displacement: The Origins of Humanitarian Emergencies – Volume 1: Analysis*, Oxford: Oxford University Press, pp. 147–79.
—— and Leach, M. 1996. *Misreading the African Landscape: Society and Ecology in a Forest Savanna Mosaic*, Cambridge: Cambridge University Press.
Fargher, B. L. 1996. *The Origins of the New Churches Movement in South Ethiopia, 1927–1944*, Leiden: Brill.
FDRE (Federal Democratic Republic of Ethiopia). 1995. *Proclamation of the Constitution of the Federal Democratic Republic of Ethiopia*, Addis Ababa: Federal Negarit.
Ferguson, J. 1990. *The Anti-Politics Machine: 'Development', Depoliticization and Bureaucratic Power in Lesotho*, Cambridge: Cambridge University Press.
——. 1999. *Expectations of Modernity: Myths and Meanings of Urban Life on the Zambian Copperbelt*, Berkeley: University of California Press.
Fernandez, J.W. (ed.). 1991. *Beyond Metaphor: The Theory of Tropes in Anthropology*, Stanford: Stanford University Press.
Fine, B. 1999. 'The Developmental State is Dead – Long Live Social Capital?',

Development and Change, 30 (1), 1–19.

Fortes, M. 1953. 'The Structure of Unilineal Descent Groups', *American Anthropology*, 55 (1): 17–41.

Fortmann, L. 1995. 'Talking Claims: Discursive Strategies in Contesting Property', *World Development*, 23 (6): 1053–63.

Freeman, D. and Pankhurst, A. (eds), 2003. *Peripheral Peoples: The Excluded Minorities of Ethiopia*, London: Hurst and Company.

Friedman, J. 1975. 'Tribes, States and Transformations', in M. Bloch (ed.), *Marxist Analyses and Social Anthropology*, New York: Malaby Press.

Fullerton Joireman, S. 1997. 'Opposition Politics and Ethnicity in Ethiopia: We Will All Go Down Together', *Journal of Modern African Studies*, 35 (3): 387-407.

Geertz, C. 1973. *The Interpretation of Cultures: Selected Essays*, New York: Basic Books Inc.

Geschiere, P. 1997. 'Kinship, Witchcraft and the Market', in R. Grinker and C. Steiner (eds), *Perspectives on Africa: A Reader in Culture, History and Representation*, Oxford: Blackwells, pp. 340–58.

Giddens, A. 1984. *The Constitution of Society: Outline of the Theory of Structuration*, Cambridge: Polity Press.

Gledhill, J. 1994. *Power and Its Disguises: Anthropological Perspectives on Politics*, London: Pluto Press.

Godelier, M. 1986. *The Mental and the Material: Thought, Economy and Society*, Thetford, Norfolk: Verso.

Grove, A.T. and Sutton, J.E.G. 1989. 'Agricultural Terracing South of the Sahara', *Azania*, XXIV: 114–22.

Guillet, D. 1987. 'Terracing and Irrigation in the Peruvian Highlands'. *Current Anthropology*, 28 (4): 409–30.

Hajer, M. 1995. *The Politics of Environmental Discourse: Ecological Modernization and the Policy Process*, Oxford: Clarendon Press.

Hakansson, T. 1989. 'Social and Political Aspects of Intensive Agriculture in East Africa', *Azania*, XXIV: 12–20.

Hallpike, C.R. 1972. *The Konso of Ethiopia: A Study of the Values of a Cushitic People*, Oxford: Clarendon Press.

——. 1968. 'The Status of Craftsmen Among the Konso of South-west Ethiopia', *Africa*, 28 (3): 258–69.

Hannerz, U. 1987 'The World In Creolisation', *Africa*, 57: 546–59.

——. 1992. *Cultural Complexity: Studies in the Social Organization of Meaning*, New York: Columbia University Press.

Hansemo Hamela, 1992. 'D'Irasha (Gidole): The Response to Social and Religious Change of a Community in Southwest Ethiopia', MA (Econ) thesis, University of Manchester.

Hardin, G. 1968. 'The Tragedy of the Commons', *Science*, 162: 1243–48.

Harriss, J. 1997. '"Missing Link" or Analytically Missing? The Concept of Social Capital', *Journal of International Development*, 9 (7): 919–71.

Harvey, D. 1989. *The Condition of Postmodernity: An Enquiry into the Origins of Cultural Change*, Oxford: Blackwell Publishers.

——. 1998. 'The Body as an Accumulation Strategy'. *Environment and Planning D: Society and Space*, 16: 401–21.

Herbert, E.W. 1993. *Iron, Gender and Power: Rituals of Transformation in African Societies*, Bloomington and Indianapolis: Indiana University Press.

Herdt, G. 1987. *The Sambia: Ritual and Gender in New Guinea*, London: Holt, Rinehart and Winston.

Hobsbawm, E. and Ranger, T. (eds) 1983. *The Invention of Tradition*, Cambridge: Canto.

Iliffe, J. 1979. *A Modern History of Tanganyika*, Cambridge: Cambridge University Press.

——. 1987. *The African Poor: A History*, Cambridge: Cambridge University Press.

Ingold, T. 2000. *The Perception of the Environment: Essays on Livelihood, Dwelling and Skill*, London: Routledge.

Jacobs, M. 1995 'Sustainable Development, Capital Substitution and Economic Humility: A Response to Beckerman'. *Environmental Values*, 4 (1): 57–68.

Jacobson-Widding, A. and van Beek, W. (eds). 1990. *Creative Communion: African Folk Models in Fertility and the Regeneration of Life*, Uppsala: Uppsala University Press.

James, W., Donham, D., Kurimoto, E. and Triulzi, A. (eds). 2002. *Remapping Ethiopia: Socialism and After*, Oxford: James Currey.

Jameson, F. 1991. *Postmodernism, or The Cultural Logic of Late Capitalism*, London: Verso.

Jensen, A.E. 1960. 'Feld-Terrassen und Megalithen', *Paideuma* 7: 258–73.

Kahn, J.S. 1981. 'Marxist Anthropology and Segmentary Societies: A review of the literature', in J.S. Kahn and J.R. Llobera (eds), *The Anthropology of Pre-capitalist Societies*, London: Macmillan, pp. 57–88.

Katz, C.1991. 'Sow What You Know: The Struggle For Social Reproduction in Rural Sudan', *Annals of the Association of American Geographers*, 81(3): 488–513.

Knutsson, K. 1967. *Authority and Change: A Study of the Kallu Institution among the Macha Galla of Ethiopia*, Goteborg: Etnologiska Studier, 28, Elanders Boktrychkeri Aktiebolag.

Kopytoff, I. 1987. 'The Internal African Frontier: The Making of African Political Culture', in I. Kopytoff (ed.), *The African Frontier: The Reproduction of Traditional African Societies*, Bloomington, Indiana: Indiana University Press, pp. 3–84.

Kruger, H-J, Berhanu Fantaw, Yonannes G-Michael and Kefeni Kajela. 1996. 'Creating an inventory of indigenous SWC measures in Ethiopia', in C. Reij, I. Scoones and C. Toulmin (eds), *Sustaining the Soil*, London: Earthscan.

Krylow, A. 1994. 'Ethnic Factors in Post-Mengistu Ethiopia', in Abebe Zegeye and S. Pausewang (eds), *Ethiopia in Change: Peasantry, Nationalism and Democracy*, London: British Academic Press, pp. 231–41.

Kurimoto, E. 2002. 'Fear and Anger: Female Versus Male Narratives Among the Anywaa', in W. James, D. Donham, E. Kurimoto and A. Triulzi (eds), *Remapping Ethiopia: Socialism and After*, Oxford: James Currey, pp. 219–38.

Lakoff, G. and Johnson, M. 1980. *Metaphors We Live By*, Chicago: University of Chicago Press.

Leach, M. and Mearns, R. (eds). 1996. *The Lie of the Land: Challenging Received Wisdom on the African Environment*, Oxford: James Currey.

Leach, M., Mearns, R. and Scoones, I. 1997. 'Environmental Entitlements: a Framework for Understanding the Institutional Dynamics of Environmental Change'. *IDS Discussion Papers*, 359: 1–39.

Lewis, H. 1982. 'Spirit Possession in Ethiopia: An Essay in Interpretation', in S. Rubenson (ed.), *Proceedings of the Seventh International Conference of Ethiopian Studies*, Addis Ababa: Institute of Ethiopian Studies; Uppsala: Scandinavian Institute of African Studies; East Lansing: Michigan State University, pp. 419–27.

Lewis, I. 1966. 'Spirit Possession and Deprivation Cults', *Man*, 1 (3): 307–29.

Li, T. M. 2000. 'Articulating Indigenous Identity in Indonesia: Resource Politics and the Tribal Slot', *Comparative Studies in Society and History*, 42 (1): 149–79.

Linares, O. 1992. *Power, Prayer and Production: the Jola of Casamance, Senegal*, Cambridge: Cambridge University Press.

Longley, C. A. 2001. 'A Social Life of Seeds: Local Management of Crop Variability in North-Western Sierra Leone'. PhD thesis, Department of Anthropology, University of London.

Lonsdale, J. 1996. 'Ethnicité morale et tribalisme politique', *Politique Africaine*, 61: 98–115.

Loomis, T. M. 2000. 'Indigenous Populations and Sustainable Development: Building on Indigenous Approaches to Holistic, Self-Determined Development', *World Development*, 28 (5): 893–910.

Lukes, S. 1986. *Power*, Oxford: Blackwell.

MacKenzie, F. 1998. *Land, Ecology and Resistance in Kenya, 1880–1952*, Edinburgh: Edinburgh University Press.

Mamdani, M. 1996. *Citizen and Subject: Contemporary Africa and the Legacy of Late Colonialism*, Princeton: Princeton University Press.

Mann, M. 1986. *The Sources of Social Power Volume 1: A History of Power from the Beginning to AD 1760*, Cambridge: Cambridge University Press.

Markakis, J. 1974. *Ethiopia: Anatomy of a Traditional Polity*, Oxford: Clarendon Press.

Masquelier, A. 1993. 'Narratives of Power, Images of Wealth: The Ritual Economy of *bori* in the Market', in J. Comaroff and J. Comaroff (eds), *Modernity and Its Malcontents: Ritual and Power in Postcolonial Africa*, Chicago: University of Chicago Press, pp. 3–33.

Mauss, M. 1990. *The Gift: The Form and Reason for Exchange in Archaic Societies*, London: Routledge.

McDowell, L. and Sharp, J. P. 1997. *Space, Gender, Knowledge: Feminist Readers*, London: Arnold.

Meillassoux, C. 1981. *Maidens Meal and Money: Capitalism and the Domestic Community*, Cambridge: Cambridge University Press.

Messeret Lejebo. 1990. 'A Study of the Technological, Social, Socio-economic and Environmental Constraints Affecting the Performance of Ethiopian Agriculture: With Special Reference to Drought and Famine Crises in Konso', *Giessener Schriften für Wirtschafts- und Regionalsociologie, Heft* 18.

Meyer, B. 1999. *Translating the Devil: Religion and Modernity among the Ewe in Ghana*, Edinburgh: Edinburgh University Press.

Miller, D. 1994. *Modernity, An Ethnographic Approach: Dualism and Mass Consumption in Trinidad*, Oxford: Berg publishers.

Mitchell, D. 2002. 'Dead Labor and the Political Economy of Landscape – California Living, California Dying', in K. Anderson, M. Domosh, S. Pile, and N. Thrift (eds), *Handbook of Cultural Geography*, London: Sage, pp. 233–48.

Moore, H. 1999. 'Gender, Symbolism and Praxis: Theoretical Approaches', in H. Moore, T. Sanders and B. Kaare (eds), *Those Who Play with Fire: Gender, Fertility and Transformation in East and Southern Africa*, London: The Athlone Press, pp. 3–37.

Murphy, W. P. and Bledsoe, C. H. 1987. 'Kinship and Territory in the History of a Kpelle Chiefdom (Liberia)', in I. Kopytoff (ed.), *The African Frontier: The Reproduction of Traditional African Societies*, Bloomington, Indiana: Indiana University Press, pp. 121–47.

Nederveen Pieterse, J. 2001. *Development Theory: Deconstructions/Reconstructions*. London: Sage.

Negussay Ayele, 1990. 'The Ethiopian Revolution: Political Aspects of the Transition from PMAC to PDRE', in M. Ottaway (ed.), *The Political Economy of Ethiopia*, New York: Praeger, pp. 11–30.

Netting, R.M. 1968. *Hill Farmers of Nigeria: Cultural Ecology of the Kofyar of the Jos Plateau*, Seattle: University of Washington Press.

Netting, R.M. 1993. *Smallholders, Householders: Farm Families and the Ecology of Intensive Sustainable Agriculture*, Stanford: Stanford University Press.

North, D.C. 1990. *Institutions, Institutional Change and Economic Performance*, Cambridge: Cambridge University Press.

Nyerges, A. Endre, 1997. 'Introduction – The Ecology of Practice', in A. Endre Nyerges (ed.), *The Ecology of Practice: Studies of Food Crop Production in Sub-Saharan West Africa*, Amsterdam: Overseas Publishers Association.

Ortner, S. 1998. 'Identities and the Hidden Life of Class', *Journal of Anthropological Research*, 54(1): 1–16.

Östberg, W. and Reij, C. 1998. 'Culture and Local Knowledge – Their Roles in Soil and Water Conservation', *Advances in GeoEcology* 31: 1349–58.

Ostrom, E. 1990. *Governing the Commons: the Evolution of Institutions for Collective Action*, Cambridge: Cambridge University Press.

Ottaway, M. and Ottaway, D. 1978. *Ethiopia: Empire in Revolution*, New York: Africana Publishing Company.

Pankhurst, H. 1992. *Gender, Development and Identity: An Ethiopian Study*, London: Zed Books.

Parsons, J. R. 1987. 'Comments', *Current Anthropology*, 28 (4) 423–4.

Pausewang S., Tronvoll, K. and Aalen, L. 2002 (eds.) *Ethiopia Since the Derg: A Decade of Democratic Pretension and Performance*, London: Zed Books.

Peet, R. and Watts, M. 1993. 'Introduction: Development Theory and Environment in an Age of Market Triumphalism', *Economic Geography*, 69: 227–53.

Putnam, R. 1993. 'The Prosperous Community: Social Capital and Public Life'. *American Prospect*, 13: 35–42.

Radcliffe, S.A.R. 2006. 'Culture in Development Thinking: Geographies, Actors and Paradigms', in S.A.R. Radcliffe (ed.), *Culture and Development in a Globalizing World: Geographies, Actors and Paradigms*, London: Routledge, pp. 1–29.

Rahnema, M. and Bawtree, V. (eds). 1997. *The Post-Development Reader*, London: Zed Books.

Ranger, T. 1983. 'The Invention of Tradition in Colonial Africa', in E. Hobsbawm and T. Ranger (eds), *The Invention of Tradition*. Cambridge: Cambridge University Press, pp. 211–62.

Reij, C., Scoones, I. and Toulmin, C. (eds). 1996. *Sustaining the Soil: Indigenous Soil and Water Conservation in Africa*, London: Earthscan.

Richards, P. 1985. *Indigenous Agricultural Revolution: Ecology and Food Production in West Africa*, London: Unwin Hyman.

——. 1986. *Coping with Hunger: Hazards and Experiments in an Upland African Rice-farming System*, London: Allen and Unwin.

——. 1988. 'Reply to Jane Guyer', *Current Anthropology*, 29 (2): 264–65.

Rowlands, M. 1987. 'Power and Moral Order in Precolonial West-Central Africa', in E. M. Brumfiel and T. K. Earl (eds), *Specialization, Exchange and Complex Societies*, Cambridge: Cambridge University Press, pp. 52–63.

Sack, R. 1986. *Human Territoriality: Its Theory and History*, Cambridge: Cambridge University Press.

Sahlins, M. 1963. 'Poor Man, Rich Man, Big-man, Chief: Political types in Melanesia and Polynesia', *Comparative Studies in Society and History*, 5: 285–303.

——. 1974. *Stone Age Economics*, London: Tavistock.

Sanders, T. 1998. 'Making Children, Making Chiefs: Gender, Power and Ritual Legitimacy', *Africa*, 68 (2): 238–62.

Sauer, C. O. 1963 [1925]. 'The Morphology of Landscape' in J. Leighly (ed.), *Land and Life: A Selection from the Writings of Carl Ortwin Sauer*, Berkeley: University of California Press, pp. 315–50.

Schech, S. and Haggis, J. 2000. *Culture and Development: A Critical Introduction*, Oxford: Blackwell.

Schlee, G. 1992. 'Ritual Topography and Ecological Use: The Gabbra of the Kenyan/Ethiopian Borderlands', in D. Parkin and E. Croll (eds), *Bush Base: Forest Farm*, London, New York: Routledge, pp. 110–128.

Schoffeleers, J. M. (ed.). 1979. *Guardians of the Land: Essays on Central African Territorial Cults*, Gweru: Mambo Press.

Schumacher, E. F. 1973. *Small is Beautiful: A Study of Economics as if People Mattered*, London: Blond and Briggs.

Scott, J.C. 1976. *The Moral Economy of the Peasant: Rebellion and Subsistence in Southeast Asia*, New Haven: Yale University Press.

——. 1985. *Weapons of the Weak: Everyday Forms of Peasant Resistance*, New Haven and London: Yale University Press.

——. 1998. *Seeing Like a State*, London and New Haven: Yale Univesity Press.

Service, E.R. 1975. *Origins of the State and Civilization*, New York: Natron.

Shako, Otto, 1994. 'The Nine Clans of Konso'. *Sociology Ethnology Bulletin*, 1 (3): 80–92.

Shinohara, T. 1993. 'The Symbolic Meaning of the Pot on Roof – a Case Study of the Konso in Southern Ethiopia', *Nilo-Ethiopian Studies*, 1: 57–74.

Shipton, P. and Goheen, M. 1992. 'Introduction: Understanding African Land-holding: Power, Wealth and Meaning', *Africa*, 62 (2): 307–25.

Sillitoe, P., Bicker, A. and Pottier, J. (eds), 2002. *Participating in Development: Approaches to Indigenous Knowledge*, London: Routledge.

Soja, E. W. 1989. *Postmodern Geographies: The Reassertion of Space in Critical Social Theory*, London: Verso.

Soper, R. 1996. 'The Nyanga Terrace Complex of Eastern Zimbabwe', *Azania*, XXXI: 1–36.

Spear, Y. and Waller, R. (eds). 1993. *Being Maasai: Ethnicity and Identity in East Africa*, London: James Currey.

Spencer, J.E. and Hale, G. A. 1961. 'The Origin, Nature and Distribution of Agricultural Terraces'. *Pacific Viewpoint*, 2: 1–38.

Sperber, D. 1975. *Rethinking Symbolism*. Cambridge: Cambridge University Press.

Sutton, J.E.G.1989. 'Towards a History of Cultivating the Fields', *Azania*, XXIV: 99–113.

Taddesse Berisso. 2002. 'Modernist Dreams and Human Suffering: Villagization among the Guji Oromo', in W. James, D. Donham, E. Kurimoto and A. Triulzi (eds), *Remapping Ethiopia: Socialism and After*, Oxford: James Currey, pp. 116–32.

Tadesse Wolde. 1992. 'The Death and Burial of Kalla Qäñazmach Kayoté, a Ritual Leader of the Konso People of Southern Ethiopia', *Sociology Ethnology Bulletin*, 1(2): 12–22.

Thesiger, W. 1987. *The Life of My Choice*, London: Collins.

Tiffen, M., Mortimore, M. and Gichuki, F. 1994. *More People, Less Erosion: Environmental Recovery in Kenya*, Chichester: Wiley.

Tilley, C. 1999. *Metaphors and Material Culture*, Oxford: Blackwell.

Turner, V.W. 1969. *The Ritual Process*, Harmondsworth: Penguin.

Turton, D. (ed.) 2006. *Ethnic Federalism: The Ethiopian Experience in Comparative Perspective*, Oxford: James Currey.

UNEUE. 2003. 'Targeting Problems Causes Malnutrition' (http://www.ochaeth.org/Archive/DownloadableReports/SNNPR0303.pdf; accessed, 4/2005).

Uphoff, N. 1996. 'Preface', in P. Blunt and D.M. Warren (eds), *Indigenous Organizations and Development*, London: IT Publications, pp. vii–xi.

Wade, R. 1988. *Village Republics: Economic Conditions for Collective Action in South India*, Cambridge: Cambridge University Press.

Ward, R.G. and Kingdon, E. 1995. *Land, Custom and Practice in South Pacific*, Cambridge: Cambridge University Press.

Warren, M. D., Slikkerveer, L. J. and Brokensha, D. (eds). 1995. *The Cultural Dimension of Development: Indigenous Knowledge Systems*, London: Intermediate Technology Publications.

Watson, E. E. 1998. 'Grand Truths: Power and Land in Konso, Ethiopia. PhD thesis, Cambridge University.

——. 2002. 'Capturing a Local Elite: The Konso Honeymoon', in W. James, D. Donham, E. Kurimoto and A. Triulzi (eds), *Remapping Ethiopia: Socialism and After*, Oxford: James Currey, pp. 198–218.

——. 2003. 'Examining the Potential of Indigenous Institutions for Development: A Perspective from Borana, Ethiopia', *Development and Change*, 35: 287–309.

——. 2004. 'What a Dolt One Is! Language Learning and Fieldwork in Geography', *Area*, 36 (1): 59–68.

—— and Lakew Regassa, 2003. 'Konso', in D. Freeman and A. Pankhurst (eds), *Peripheral People: The Excluded Minorities of Ethiopia*, London: C. Hurst and Co., pp. 240–61.

——. 2006. 'Culture and Conservation in Post-conflict Africa', in S.A.R. Radcliffe (ed.), *Culture and Development in a Globalizing World: Geographies, Actors and Paradigms*, London: Routledge, pp. 58–82.

Watson, E.E., Black, R. and Harrison, E. 1999. 'Reconstruction of Natural Resource Management Institutions: A framework for research', *Marena Research Project Working Paper, University of Sussex*, 1: 1–23. (http://www.geog.sussex.ac.uk/research/development/marena/pdf/wp1.pdf)

Weber, M. 1978. *Economy and Society: An Outline of Interpretative Sociology. Volume 1*, Berkeley: University of California Press.

Widgren, M. 2004. 'Towards a Historical Geography of Intensive Farming in Eastern Africa', in M. Widgren and J.E.G. Sutton (eds), *Islands of Intensive Agriculture in Eastern Africa: Past and Present*, Oxford: James Currey, pp. 1–18.

—— and Sutton, J.E.G. (eds). 2004. *Islands of Intensive Agriculture in Eastern Africa: Past and Present*, Oxford: James Currey.

Wittfogel, K. 1957. *Oriental Despotism*, New Haven: Yale University Press.

Wood, J.C. 1999. *When Men Are Women: Manhood Among Gabra Nomads of East Africa*, Madison, Wisconsin: University of Wisconsin Press.

Yeraswork Admassie. 2000. *Twenty Years to Nowhere : Property Rights, Land Management and Conservation in Ethiopia*, Lawrenceville: Red Sea Press.

Young, J. 1996. 'Ethnicity and Power in Ethiopia', *Review of African Political Economy*, 70: 531–42.

Index

Printed and bound by CPI Group (UK) Ltd, Croydon, CR0 4YY

16/09/2026

14963022-0001